HORSE GIRLS

HORSE GIRLS

RECOVERING,
ASPIRING, AND
DEVOTED RIDERS
REDEFINE THE
ICONIC BOND

EDITED BY
HALIMAH MARCUS

HARPER ⬤ PERENNIAL

NEW YORK • LONDON • TORONTO • SYDNEY • NEW DELHI • AUCKLAND

The names of some individuals mentioned in this book have been changed to protect their privacy.

HARPER ● PERENNIAL

HarperCollins books may be purchased for educational, business, or sales promotional use. For information, please email the Special Markets Department at SPsales@harpercollins.com.

FIRST EDITION

Designed by Jen Overstreet

Library of Congress Cataloging-in-Publication Data has been applied for.

ISBN 978-0-06-300925-7

21 22 23 24 25   BR   10 9 8 7 6 5 4 3 2 1

For the women who taught me to ride

She had some horses she loved.
She had some horses she hated.
These were the same horses.

—JOY HARJO

CONTENTS

INTRODUCTION
HALIMAH MARCUS

This past January, as my New Year's resolution, I started taking riding lessons again. The farm I've chosen is on a winding country road tracked with red clay from the unpaved driveways. There, I am given a tour and instructions to tack up Woody, the seventeen-hand paint I will ride in my lesson. I'm shown the tack room, with hooks for cleaning bridles, saddle pads stacked a dozen high, and the office, walls lined with faded ribbons and dusty trophies. Every barn I've been to has some version of these spaces, each unique in its particulars, but sensorially the same.

In one hand I carry the green plastic tack box I got for my eleventh birthday, and under my arm is the brimless crash helmet I wore when I galloped cross-country as a teenager. It has been fifteen years since I've ridden regularly, ten since my last lesson, but the routines return to me easily. There's a hierarchy at all barns: the trainer (Debbie, in this case) is at the top, followed by the most dedicated riders, who tend to have the most expensive horses. Riders who neither lease nor own a horse are near the bottom. These are the riders who mount "lesson horses," which are usually some combination of old, fat, and slow. A riding friend of mine used to say these horses were "dead to the world," because nothing startles them. They are dependable, they won't shy at an open door, they'll let you touch their faces and ears and girth. Often, there are stories about these horses' glory days. Woody, the horse I'm riding, is twenty-one but used to be a star show jumper, I'm told. Now he leads an easier life carrying inexperienced—or, in my case, formerly experienced—riders around the indoor ring.

At my first lesson it's very important to me that I let everyone know how long it's been, exactly how good I used to be. I did the whole thing: competitions every weekend; three-foot-six fences, ditches, drops, and water jumps; polished leather and pressed stock ties; dressage tests and leg yields. I still have my seat—straight spine, loose hips, shoulders square—but the obscure muscles on the insides of my thighs haven't been properly exercised in years. I struggle to make Woody go. Our trot has no impulsion; he breaks from the canter. Debbie's firm commands are comforting in their familiarity. Hands higher, softer. Toes in, heels down. She identifies the bad habits that used to plague me in my eventing days—a hip that tilts in around turns, a seat that drives when it should be light. Debbie self-describes as tough. "I don't baby people," she tells me, but I wouldn't have wanted her to. Fifteen minutes in and already I feel myself returning to the star pupil I used to be, lengthening my spine, answering yes instead of yeah. "Am I boring you?" my former trainer used to say if I ever yawned in her presence, even when we gathered at the barn in the purple predawn the morning of a show.

I know how to elicit and appreciate love from women like this. Strong, independent businesswomen who demonstrate affection by raising jumps and taking away your stirrups. By inviting you to ride in the truck with them or giving you a private lesson during an unexpectedly free hour. By pushing you, which, if you speak their language, means they believe in you. By keeping expectations always one notch higher than the level where you are.

After the lesson, I groom Woody on the cross ties as teenage riders dart around—horse girls, one might call them, because they ride horses and they are girls. I recognize myself in them right away. They are in their element, laughing and joking, unburdened by whoever they are at home or at school.

I easily identify the top ranking among them, who has curried the coat of her black Thoroughbred to gleam. She gets her friend to record a TikTok and lifts her horse's upper lip so the pink, glossy underside is revealed, the place where off-track Thoroughbreds hide their racing tattoos, and makes him talk, doing his voice. My horse friends and I used to carry disposable cameras to the barn, purchased expensive prints from the show photographers for our carefully curated photo albums, or sometimes, settled for the wallet-size proofs with their watermarks and punched holes. I want to tell her this, but I know she won't care. I am twice her age, and I am riding a lesson horse.

I was once a horse girl, but I never became a horsewoman. Horsewomen are tough, no nonsense, fit but weathered, usually with a bad knee or some old injury acting up, but still always up early, riding every day—birthdays, New Year's, in snowstorms, and on Christmas. Unsentimental about horses but devoted to them for life. Women like Debbie. But not me.

Like many of the writers in this anthology, I stopped riding when I went to college, quitting the sport at the final edge of girlhood. I look back on it now as inevitable, the forces of money, time, and education conspiring against me, but then it felt like a crossroads. I could either attend a local school, maybe even major in equine studies at one of the rural institutions, or I could pursue a liberal arts degree at a more demanding and prestigious university. The former option would have allowed me to live at home and continue to ride competitively, working toward a professional career. The latter would be the first step on a path to pursue my quieter dream to become a writer.

In her essay "I Don't Love Horses," T Kira Madden writes cannily how, because she is a writer, people often mishear her recounting of a riding accident as *writing* accident. It's funny to imagine a writing accident that would land someone in a neck brace, but the cognitive dissonance of their reactions also shows how rarely these two worlds overlap. As I approached the end of high school, I believed that if I became a rider, I would never be a writer. Both because I wouldn't be educated as a writer, and because the world of horses, as I knew it, had nothing in common with the world of arts and letters. If I continued to ride, I not only wouldn't be a writer, I wouldn't know any either.

This either/or dichotomy, like most dichotomies, was based on a series of unexplored assumptions. Of course there are riders who are also writers—I found fourteen of them for this book, and there are many more I wish I could have included. Back then, as a senior in high school, I chalked the choice up to logistics—how would I ride if I was busy at school, how would I get a good education if I was busy riding—when really, it was based on a more insidious assumption that if I rode, my life would lack creativity and inspiration. If I rode, I wouldn't have anything to write about.

Wanting to make a clean break, I applied early decision to an elite liberal arts school that did not have an equestrian team, and got in. My parents made plans to lease my horse, Dave, so that I could come home and ride him on holidays and weekends.

On the first day of a writing class freshman year, we were asked to introduce ourselves by providing a unique piece of trivia. "I used to be heavily into riding," I said when it was my turn, though I had quit practically yesterday and technically still owned a horse. Only as the words came out did I realize that I sounded like a snob, that the word "heavily" was awkward, menstrual, unintelligent. My classmates looked at me blankly. "So am I!" chimed another girl. Already I could tell she was friendly and enthusiastic and had no sense of irony. A full-on horse girl. I'd admitted the same about myself just a few moments earlier, but at least I had the good sense to be embarrassed by it. Later, she joined the college's extracurricular riding club and proudly strode across campus in her jodhpurs, wrote stories (in college!) about sentimental relationships with horses while the rest of us tried to find a "voice" as serious writers. My school was nearly five hours away from home by train. Dave was sold by Christmas.

Over the next four years, I reinvented myself as anything but a horse girl. I played indie music on the college radio station, dressed in vintage clothes, and wrote moody short stories that didn't feature any animals, let alone horses. Horse girls were unfashionable, out of touch, unsophisticated. My new hobbies were obscure and exclusive, my demeanor disaffected. My assumption that horses could never be subjects of literary merit remained unchallenged. In fact, I didn't believe that anything in my life was worth writing about. I made fun of classmates who set their stories in dormitories and on the quad, in their childhood bedrooms and high school hallways. I wrote stories that had nothing to do with me, that blatantly imitated the type of work we were assigned in class, stories about stoic longshoremen, single mothers driving cross-country, and alcoholic brides getting cold feet. Yet I had no interest in inventing a cowboy with a trusty mustang, or an heiress with a stable full of racehorses, a lonely farm girl with a stubborn mare. When I tried to push my imagination beyond imitation, all I got were more clichés.

I used these stories to get into graduate school in New York, an MFA program in fiction, where I by now had some idea about how to fit in with the urbane, liberal, and artistic individuals I wanted to be like. My writing grew more personal, but riding remained a source of shame, something I rarely mentioned, or talked about only jokingly. It seemed aristocratic and conservative, a bucolic fantasy that had no place in New York City. And yet, I pined after it privately. From my life in the city, this pining was necessarily done at a remove. My Instagram feed filled up with horses, I watched livestreams of competitions on the weekends, I had a recurring fantasy that a lesson horse in Prospect Park would get loose during one of my regular runs, and I would dash valiantly to the rescue. (Unlike a doctor, who might reasonably be called upon at any moment, *Does anyone know how to ride a horse?* is not a question I was likely to encounter anywhere, for any reason.) Instead of growing out of my love for horses, as I always assumed I would, I felt that horses were drawing closer, demanding more attention, becoming more insistent in their absence.

I confided in my new boyfriend, whom I met in the writing program, about how much I missed riding. Though we'd only been dating a few months, he arranged for me to have a lesson at a farm in New Jersey for my twenty-fifth birthday. "I thought you'd be wearing the outfit," he said, disappointed, when he saw me dressed in the navy-blue riding pants, paddock boots, and half chaps my mom had mailed from Pennsylvania for the occasion. "You know, those tan leggings and tall boots."

I knew the look, the magazine-perfect horse girl he imagined, but was still surprised to hear him desire her so bluntly. As a girl, I'd coveted that image in an asexual way, wanting the perfection, the thinness. Instead of questioning what it was about the equestrian uniform—prim and prescribed, like a Catholic schoolgirl's—that he found so alluring, I thought instead that there was no way my tall boots, which I hadn't worn in years, would fit now. (C. Morgan Babst writes poignantly about the symbolism of that too-slim leather, straining against calves that have grown too wide.) Just the mention of those boots recalled the compression socks and the baby powder, the boot pulls and the bootjack, the phantom of charley horses past, a display I could no longer muster.

The lesson was at a dressage stable helmed by a former Olympian, and the horse they put me on, while technically a lesson horse, could still do piaffes and passages, flying lead changes, and extended trots. I had never ridden a horse purely trained for dressage; the Thoroughbreds I evented in Pennsylvania sacrificed flare for power. During the lesson I felt that feeling riders chase: complete syncopation with my steed, a half ton of muscle and flesh floating above the ground. After, sweaty and proud, I asked my boyfriend what he thought. My beaming face anticipated the praise I was about to receive. He told me he hadn't realized he was supposed to watch, as if the outfit was the only show; he'd spent the whole time waiting in the car.

Still, I told friends the lesson was "the best present anyone has ever given me," and interpreted it as a sign that this boyfriend really knew me, that he saw value in my dreams, even the dream I was embarrassed to admit to, the one I'd permanently deferred. I realized I'd given him too much credit when he later reported to me a "hilarious" conversation he had with our classmate about "how weird horse girls are." He told me this without any inclination that I would be insulted, as if I would agree enthusiastically, as if even as the alleged horse girl, I could still be in on the joke.

The boyfriend and I didn't last, and these slights were, admittedly, minor. I consider it a flaw in my character that, ever since I was one of those weird horse girls myself, I have been susceptible to an eye roll, to a laugh at my expense. Writing this, I'm even embarrassed to admit how long that moment has stayed with me; I should have forgotten it years ago. But there's no ignoring that I waited ten years to have another lesson, until the day I rode Woody. My ex-boyfriend's comment tapped into an insecurity, a suspicion that the sport I loved limited me to an array of contradictory clichés: an overzealous misfit; a girl with her head in the pastures; a princess devoted to ponies and privilege; a dominatrix with high boots and crop.

Unlike other sports, which are as ingrained into our national consciousness as the seasons, competitive riding inspires only other riders. It doesn't foster civic pride, or cross-cultural conversation. It doesn't give you something to talk about with your neighbor, or your coworker, or your dad. Outside of tight-knit riding communities, its value is primarily personal, deriving from the bond between rider and horse, a bond that can feel ancient, even spiritual, but is nonetheless about subordination, a bond made possible by centuries of domestication.

One weekend a few years ago, I retreated to a friend's cabin to work on an assignment that would eventually inspire this anthology. My longing to ride, but inability to fit it into my life, had finally taken up enough space in my mind that I had no choice but to write about it. It was the first time I'd attempted to write about horses, and everything I came up with felt myopic and indulgent. I left my desk and went for a walk in the woods, following whatever path called to me, uncharacteristically without a map or a plan. I headed down one trail, hoping it would loop back around, but as it kept going a feeling pestered me: if I continued, I would soon be lost. When I turned to retrace my steps, I saw, spray-painted on a boulder, "When we ride a horse we borrow freedom."

As far as I knew, riders did not use this park. This was not horse country. The message had been at my back, and had I not turned around, I never would have seen it. The message had been written for me.

Freedom, yes. Freedom was the feeling I got from riding, though I would have been too self-conscious to put it so plainly. Galloping, all four legs off the ground, six if you count mine. People can't help but compare it to flying. And that word: borrow. The best horse and rider relationships have a give-and-take, their own language, a physical communication. When we ride a horse we don't take their freedom; their freedom is not lost by being ridden. We borrow it; they share it with us; it is not depleted by being shared, and it is endlessly replenished. Freedom is not bestowed upon a horse; it is an innate quality that horses, even in captivity, embody and exude. Horses are free in ways that humans can only sample. We do not permit horses to be free because that kind of freedom is not a gift that humans have to give.

Riding freed me from myself, but only temporarily. Self-consciousness returned as soon as my feet touched the ground. The way I continued to long for horses as an adult forced me to reckon with why, in the first eighteen years of my life, my sense of self was so dependent on something I was determined to deny in the next eighteen. Rather than view it as straightforward and sentimental, I began to see the damage of the horse girl stereotype, from the strange way that so-called horse girls have been both fetishized and made fun of in our culture, infantilized, sexualized, and mocked, to the way that the very term negates the athleticism and bravery required to ride a horse. Though it's a label few apply to themselves, I also thought about who the term *horse girl* excludes: anyone on the margins. As Carmen Maria Machado writes in "Horse Girl: An Inquiry," "If you were to lean close and breathe deep, she would smell like heterosexuality, independence, whiteness, femininity."

As I edited the essays that would be in this collection, I recognized certain patterns. Shame and passion are in conflict in many of these stories, albeit from different sources. Shame over not looking the right way, over not having enough money, of being too chubby, or too boyish, or queer, or having dark skin. Perhaps shame is a defining characteristic of girlhood, a period that concerns most of the essays here. It would follow, then, that girls who experience acute shame are motivated to seek relief for it on horseback. Riding promised a way to escape our circumstances and our bodies, our genders, our parents, our upbringings. On horseback, we weren't horses, but we weren't girls, either.

The average life span of a horse is twenty-five to thirty-three years. Their prime ages for racing are between seven and ten, eleven to seventeen for other disciplines, like show jumping and eventing. An eighteen-year-old dressage horse is an elder statesman. Horses age much like humans; their hair becomes flecked with gray, their faces hollow. Concavities appear above their eyes, rather than at their cheeks. Their joints ache, they grow stiff. When they sleep, their lips hang slack.

At thirty-five, I am at the upper end of a horse's life span—decidedly a woman, no longer a girl. Finally mature enough to be in on, or at least on the outside of, the horse girl joke. There wasn't a precise moment when I stopped being a girl and became a woman, but, while I haven't been a girl a long time, there were many years when I was both. Nur Nasreen Ibrahim calls the transition between girlhood and womanhood "small blasts of understanding." Maggie Shipstead writes of a gradual epiphany. Alex Marzano-Lesnevich imagines the girl they once were as an "artifact of history."

Most of the contributors to this collection stopped riding at some point, as I did, usually in their teenage years. For many in adulthood, riding indeed becomes a series of personal artifacts, a closed photo album, a box of ribbons, boots that no longer fit, a leather halter drying out. Some started riding again in their twenties, thirties, or forties, after marriages and divorces, trips around the world, bouts of depression, career changes, and children. They write about how growing up changed them, of what was lost: courage, or maybe recklessness, uncomplicated relationships with horses that are founded in innocence, the joy that comes from doing something just because you feel like it, the privilege of not asking questions.

In my riding heyday, my horse Dave and I were the same age. My parents bought him when I was sixteen, when I was rising to the highest ranks I would ever achieve in equestrian competition, and he, unbeknownst to me, had just passed his. That we were the same age felt at once cosmically significant but also like a bit of novel trivia. At sixteen, I was checking off milestones, doing all the things teenagers do to grow up, but I was still a young rider. At the same age, Dave was an experienced competitor, had placed in the top five at high-level

Three-Day Events. I was his second or third or maybe fourth owner; he had already brought one of the older girls at the barn up through the levels. Predictably, she sold him to me when she went off to college.

If Dave were still alive today, he'd be elderly. I kept tabs on him for a while, watching on Facebook as he taught girls even younger than I was how to ride, carrying them over fences at their first competitions. To Dave, horse girls were a renewable resource. We grow up, and we quit, we leave our horses behind, but the horses don't care. Dave went from being *my* horse to being a lesson horse, shared by many but belonging to no one.

Any rider who doesn't quit, who stays in it for their lifetime, will outlive their horse. Adrienne Celt writes about worrying about the eventual death of her horse on the day she bought it; a day that should have been happy but was instead spent anticipating future grief. Jane Smiley, who returned to riding as an adult and spent years breeding racehorses, has seen horses through their entire lives, from foalhood to pasture. A few years ago, when I heard that Dave had died, I cried as I might have done if it had happened when I was young, if he had died when I still knew him. Though I'd long since grown up and moved on, though it'd been years since I'd last felt the weight of one of his hooves in my palm, since I'd pressed my forehead into the base of his neck, since I'd watched him gleefully cake his freshly bathed coat in mud. Losing him made me a girl again.

When I returned to riding as an adult I no longer had the run of the place. The girls whose voices echoed down the concrete aisle, who called to one another between stalls, were not my friends, or my peers. There wasn't a horse, which I had seen yesterday and would see again tomorrow, waiting for me in the field. Instead there was just my name misspelled on the lesson board, next to the name of a horse I would ride for the first, but maybe not the last, time.

When does a girl become a woman? Is it when she turns eighteen and becomes a legal adult? Is it when she loses her virginity, or when she first experiences a loss that forces maturity, or a trauma that shatters her naive sense of invincibility? Is it when she gets her first job, or falls in love, or gets married, or has children? Is it when she understands she was never a girl to begin with, and she instead becomes he, or they? When does a horse girl get to become a horsewoman, a horse person?

Not all horse girls must leave horses behind in order to grow up. There's nothing about riding that is innately infantilizing, or immature, or feminine. Because it can be understood in human years, perhaps a horse's life span also describes the life span of girlhood, from its first inklings to its last gasp. We cling to the child within us—by reliving our past glory, by taking our hobbies seriously, by identifying with old insecurities—and we fight it, by putting ourselves at risk, by taking on more than we can handle, by denying our simple passions. But all that thrashing does nothing to forestall the natural way of things. A horse lives three decades, from rambunctious colt to seasoned athlete to lesson horse. The older the horse, the younger, or more inexperienced, the rider.

By returning to a lesson horse, I was humbled, forced to admit that I'd given up my place, that I still had a lot to learn. I was also compelled to admit that the divisions between the two sides of myself, which I had previously considered irreconcilable, were of my own invention. As riders, as humans, we must take the lessons from horses that we can. If I could be an adult and ride a lesson horse, if I could be experienced and still be a beginner, maybe it was possible to love horses, to express that love in all its contradictions and nuance, and enjoy the ride.

I DON'T LOVE HORSES
T KIRA MADDEN

Sonora Webster, or Gabrielle Anwar playing the real-life Sonora Webster Carver in the Disney movie *Wild Hearts Can't
Be Broken*, is so hot. That's how it feels when I'm a kid. I'm a
kid in a room with a TV, right up close to the screen, and kid
me doesn't know she's gay yet. She loves Sonora especially
when she scissors off her own hair, and even more when Sonora jumps on her horse, Lightning, in a field, bareback, no
tack. No fear. She falls, of course. She doesn't get good until
later in the movie.

When I am twenty-nine years old, I, too, mount a horse
in my girlfriend's backyard. That horse—Hank—did not look
unlike Lightning. No tack, a few drinks in me. He looked so
kind, so ready for it. Of course, I fall. Sometimes to build a
connection is to build a cliché. Hank didn't throw me, it was
a tree. There were other horses in that backyard paddock, and
those horses chased Hank into the woods. I had no reins; a
tree branch caught me in the head; wipeout. But the details
don't matter, I don't matter. I am just another horse accident.

Riding accident, I told people after, when they asked about
my neck brace.

They heard *writing accident*. A neighbor asked how I split
my head on a desk.

No, I *ride*.

And then, the usual: *But what is riding really | I mean it's not actually a sport not like a football sport | I used to ride as a kid | I rode once no I don't remember English or Western | Don't you just kind of hold on and steer | Doesn't the horse do all the work?*

The fall was my fault. I was drunk, remember. It was all very boring, but now it's a story. There was something so *Sonora* about it—something *Gabrielle Anwar playing Sonora in a Disney movie* about it—that makes the whole crash, now, feel mighty, feel Wild Girl. I've tried to write about horses so many times. But the thing about a horse is, it's never about the horse.

Everyone has a horse story.

Horse people know this.

The moment you announce yourself, bring your old show days or horse anecdote or *writing accident* into a room, you'll get The Horse Story. Someone was thrown by a Bad Horse, once. They never got on again. Another horse tore across a field—*He was crazy! Like a mustang!*—but the storyteller held on heroically. Another woman, one you meet at a wedding, says she used to groom horses at summer camp and still misses those "big, fluffy paws" they had. That horse *trusted* her, she said. *Intuition*—the pawed horse could feel her goodness.

Did you bleed the first time? she asks. *Like, down there?*

Horse as pawed, intuitive being becomes the Hymen-Breaking Horse.

Then comes another story, from someone else.

Well, that sounds special, is how it goes.

———

Here's something horse people do: there is always The Age at which you first got on, gripped a saddle horn or a mane. *By the time I could walk*, or, *Ever since my first pony ride*. My mom tells it that I was two, which is about standard as far as The Age goes. A trail ride in North Carolina—the smile on my face—*everybody knew*. But I don't remember joy as much as I remember balance, twitching ears, the boniness of withers, the power of steering. How, with those reins in my hands, I was not being led by a stroller, a tug at the arm, the clutch of an adult.

I dreamed as a child of screwing up that steering, all that responsibility, jerking my reins more to one side than the other; I couldn't control those yanks. A bent horse neck, never centered. Then, a fall right over the horse's body. I lived in a storm of a household and developed early obsessive-compulsive disorder, ritualizing symmetry.

Horses will help, everybody said.

Everyone knows horses are therapy.

I jerked and jerked in those dreams. Every fall more brutal. Trampled.

I keep soft hands these days. Barely touch the reins. One of my favorite exercises, still, is arms out, no steering. It was one of the first things my wife ever noticed about me. But that was later, thirteen years after *never again*, when I finally got back on.

Horses, they're so goddamn American. A man is made sexy in proximity to his horse, his seat atop a beating-hearted animal. Wealth, status, that shining coat of power, a beast "broken" by its master. Broken. Saddle broke. To break green. Green broke. Those are the terms for a horse tamed, *bettered*, accepting this new weight on its back.

I turn on the television. A bay, lunged by a small, pig-tailed brat. The brat knows what she's doing, shows the horse who's boss in this relationship. The horse whinnies to no one, just to announce its horseness. Of course it does. There's always the announcement, the stock sound.

There is the white horse. The black stallion. The biblical horse and the horses of war. The girl who learns to love again because *Horse!* and the gruff man who softens, heals, in the presence of all that Horse Majesty. There's the underdog horse and the horse who brings the family together in unexpected ways; a child, free with gladness, atop their first rocking horse. There's the Horse to the Rescue and the horse who knows *exactly which way to go*. The horse that falls off the cliff; horses, our saviors, collapsed in battle. What they don't tell you about are the trip wires that made the shots.

Recently, someone brings up National Velvet, the horse.

The horse wasn't Velvet, I say. *Elizabeth Taylor was.*

The horse was Pie.

I don't love horses. But of course I do. I did as a child, and as a teenager, and as an adult, and especially now. Sometimes it feels as if I was meant to love them, and I am possessive in that love. When I am in the presence of the basic Horse Story, a rage thickens in me. *You don't deserve that story*, I think. *You don't even know.* But, of course, my love, too, is the problem. How many times were my horses there to deliver me to a dream, a goal, a peace. *Everybody knew.* I'm a teacher these days, and I tell students often: *if there's a problem, write into the problem.*

I was a hunter jumper first. Then, one summer barrel racing. Then, a few years training to be a jockey. There were so many outfits, personas. So many shows. Numbers strapped to my back, embroidered saddle pads, custom chaps, helmet silks—new equipment for every new discipline. I still wear the gold nameplate leather bracelet with the name of my first pony, Cloud 9, stamped in. He's a symbol now, too. *He saved my life,* I tell people. *I could tell the weather by the temperature of his nose,* I say. *My first best friend,* I wrote in the acknowledgments of my first book. We never made it big. There was no hero's journey. My parents gave him away, and I never asked after him, simply because I grew up, and I took things for granted (that pigtailed brat), and I think I was afraid of where he went because it meant he was no longer mine.

I reunited with him when I was twenty-five and he was thirty-two, for one day, and I like to think he remembered me, though this, too, could be romantic thinking. He died soon after. I waited for his horseshoes in the mail, but they never came.

Nicky was his barn name. A Welsh pony who was always hungry, with the markings of a heart on his rump. He was so stubborn and so good. He knew how to give hugs by curling his neck around you. He loved apple pie and once he threw me so he could run to the show grounds hot dog stand.

I want to write about the horses, not just the rider. I want to find where the two break.

Gabrielle Anwar playing Sonora Webster strokes her horse Lightning in *Wild Hearts Can't Be Broken* almost erotically. She's lost her sight by this part of the movie, by this point in her life—detached retinas from an open-eyed dive—so she's feeling around, getting a grip of it all, as music thumps in the background of the scene, a storm brewing—cue *lightning* flash!

Lightning, a gelding in the movie, will deliver Sonora to her destination of morality, growth, glory, a standing ovation. Good boy.

What the Disney movie leaves out is how the real-life Lightning, a mare, was later forced to dive into the Pacific Ocean, instead of a water tank, for a California diving show. Confused by the currents, the pier, the rough breakers, the horse swam out into the open sea. Kept swimming until no one could reach her. Until her drowned body was towed back to shore by a single rope.

Sonora wasn't even there.

Luhi ʻuʻa i ka ʻai a ka lio is a Hawaiian proverb translated by Mary Kawena Pukui as "wasted time and labor getting food for the horse." The rough breakdown of its meaning goes like this: if a person works hard, wins big, and brings in money and rewards to share with friends, those friends will use it up, take it all, and then move on to the next person who might be so generous, a person who has *more to give*. I've read this proverb many times, at first thinking the horse was a stand-in for the greedy friends, endlessly fed. Then I read the reversal: the horse as the subject, the one who gives and brings it all, sacrifices it all. And we take take take take, of course we do. But maybe the horse is not the hard worker, or the friend. Maybe the horse is simply the horse, the prop, the superfluous detail used to make the point.

I had another horse. Bidster, my old chestnut Thoroughbred, and he's still alive. I was there when he was born, when I was around seven years old, but my memory of that night is spotty. Slimed webs of membrane, all legs, a damp barn lit by orange light. I called him Glitter Man when he was born like that—wet and glowing for the world—which is still his show name. He felt like mine.

Bidster belonged to a man named Frank, a battle-worn jockey who'd been rail pinned and trampled on the track, leaving him with a limp. I worked for Frank and his wife in the summers, helping them tack up for trail rides, mucking stalls. When Bidster was born, Frank said we were the winning ticket. His way back to the tracks. So we broke him together. We found open fields and construction sites in Seven Devils, North Carolina—the dead maw of open cranes, towers of bricks—land, I'd learn later, that did not belong to me. Cherokee land. Burial land. Which, back then, to kid me, carried a spooky importance. When I think about breaking a racehorse, tightening the girth, smacking that crop, on land that felt so much like *ours*, there is only shame there. The me of now would tell kid me this: there are always more important stories beneath your stories. Nothing is yours.

I wore a helmet with green-and-white jockey silks over it, and Frank threw dirt from behind us: *I want your hands at his ears, that's how far forward*. It is true that we were good together.

Frank moved to Florida and into our house so we could train every day. We watched what felt like every race ever raced on my father's big-screen TV. We went to Gulfstream. We studied stats, and I learned how to read them. Frank and his newspaper hands, blackened at the fingertips. I was good, he said. Everybody said it.

Then, my body bulged. Puberty. Et cetera. You know this story.

When we failed as a racing pair, Frank and I took Bid to a show in Yadkinville, North Carolina. This time, a Jumper division. He spooked; he refused; he hated crowds; we didn't clear a single fence; my posture curled like burning wood; we were pinned dead last. I never showed again.

Glitter Man is a "pony horse" now, meaning he escorts the star jockeys and mounts to the gates at Belmont. I went to see him not long ago and, you know, I'd like to think he remembered me.

Just last month, Frank called to tell me Bid's owner was looking to give him away.

Doesn't the horse do all the work?

Gabrielle's Sonora wears a paper bag over her head after that haircut, after her aunt (in real life, her mother) suggests she ought to be ashamed of her ugliness. Sonora leaves the house anyways, stomps the paper bag into the ground. I've never worn a bag over my head, although, on several occasions, I have wanted to. Real Sonora let her boss, Doc Carver, tell her what to do with her hair and how to dress for the rest of her life, even after he died. To be a diving girl, you must play the role.

In the final movie dive, Gabrielle Anwar's Sonora Webster reaches her hands out for Lightning and listens to every amplified stomp of those hooves up the ramp; the audience gets that slow-motion listening, too. Gabrielle's Sonora has been practicing for this moment; she's trained and ready; we are all in on it. We all get the applause.

Real Sonora always wore a helmet after her accident; it was custom-made with a special shield to protect her eyes, just in case any future medical advancements could restore her vision. Real Sonora did not practice. Her first go at diving without sight happened because the other diving girl canceled. She pulled on her special helmet and waited at the top of the ramp, only to realize the helmet made it impossible for her to hear. There was no stomping, no dramatic countdown. It was more amazing than that, more amazing than the movie. She reached her hands out and just, quite simply, *knew*.

When the fabled, stand-in, whinnying horse needs more impact, horses are anthropomorphized. The high-pitched voice-over in *Black Beauty*. Mister Ed, with a thread in his mouth to cue the movements of his talking.

Beauty, played by Docs Keepin Time in the 1994 film, is mercilessly abused, disfigured, sold, repeatedly forgotten. The voice-over narrates, tells us just how bad it is. In the *Entertainment Weekly* review of *Black Beauty*, Lisa Schwarzbaum writes, "Girls will inevitably love all this. Boys will torment those girls by saying 'oats, oats, oats!' in twitty voices that make their sisters cry."

There were two Mister Eds. One: the acting horse, a gelding named Bamboo Harvester. Another palomino named Punkin was used only for photographs and press. The former horse was, according to some accounts, accidentally murdered by "inadvertent tranquilizer." The horses became interchangeable, one replaced the other, and still there is only a single grave marker, shared by them both. The granite tombstone does not feature either horse's name; it reads, simply, "Mister Ed."

When Al Carver, the real Sonora Webster's husband, was questioned by the S.P.C.A., pausing their diving act, Al loaded one of their horses onto the bed of a truck with a sign that read *I'm being taken to jail for jumping in a tank of water!* He drove that truck all over town. He then brought Lightning to the courthouse and made the judge walk outside, take a look at this poor, beautiful girl, useless without her job, and just like that, they were back in business. *Doesn't the horse do all the work?*

There is thrill in trying to re-create history, and I am troubled when my body halts that re-creation. I try to fit into the old clothes—of course not. I order the same blue ProStretch tool to stretch my calves by rocking, readying myself for the saddle. The tool worked then, but nothing now. Overdeveloped hamstrings, no give, taut mess of muscles. Still, the blue of it in the corner of my room offers a comfort.

My velvet helmet, too, was laughed down when I showed up to ride in it thirteen years after I quit. Not up to snuff, that thin pathetic shell, phased out years ago by new safety standards. My once-gleaming show helmet, the blackest of black, had marooned in the sun. It sits on a bookshelf now. Funny. Decorative.

The question comes up sometimes, still. *Why did you stop?* But that answer is simple. I'm more interested in why I started again.

My wife says I'm one of the only riders she knows who still rides with joy. By that, I think she means I'm not as burdened by what it all means—the barns, the aspiring show girls applying glue to the insides of their boots, the privileged elitism, the horse doping, the abusers, all the horses who are "given away," disappeared. The undocumented grooms—braiding those manes, polishing leather, offering a leg up to the next pigtailed brat—who, after horse show sweeps, also disappear. I got out of the industry early enough, perhaps, to ride past that.

I am still riding for the horse of it.

Or, I am trying to.

Sonora, I think, tried, too. "The drop from the tower down to the tank is a pleasure totally lacking in psychological or philosophical meaning. It's the sheer exhilaration of being entirely free of the earth as well as everything human; to me no other physical sensation can be so acute, so deeply intoxicating," she wrote.

Free of the earth.

But for whom?

No, I've always said, *of course the horse doesn't do all the work*. The rider does plenty. It's the rider memorizing the course, the surroundings, the footing; the rider directing; the rider working her body *just as hard* to apply pressure and then to release; the rider is counting, the rider is steering; the rider has shined her boots, she's been practicing, she knows enough to know when Lightning is right *there* without seeing or hearing. The rider has done so much to win.

Hank has died by now. Nicky has died by now. They are both buried in backyards, where other horses roam. I tell Frank I'll take Bidster in, my old retiree, because I'd rather pay his medical bills than have him disappear to *who knows where*. My other ponies disappeared to who knows where; I never asked. I'd left those ponies out of the story.

I still have the halter, bright pink with puffy paint, of my miniature horse named Tulip. The puffy paint is that of my own, shaky child hand. Tulip, it says, with drawings of not-Tulip flowers. Decorative now, too. Hanging in the guest bedroom, with my spurs, just for show.

I have one photograph of my bay pony, who my mother and I saved at auction for $200. She was always sick and couldn't quite recover, and the photograph shows it—swollen eyes, a wet nose, emaciated. I broke her, but didn't have her long. Her name was not Pie. I named her Velvet.

Maybe I ride with joy because I'm able to imitate some of my own Horse Stories, and all the others out there. Because I'm Sonora and I'm Velvet. Because I'm still me, the kid, who once loved horses before I swore I didn't and then I did again. Who once thought an animal could be broken, owned, mine. Now I want so badly not to love them, to step out of the problem. If I'm not the brat in pigtails perhaps I could be a Noble Horsewoman, reclaiming something by dropping the reins, letting go.

When I sold my first book, I saved a small portion of the advance money to buy myself a saddle. I returned to the HITS show grounds to pick one out; it was used and soft, my first adult saddle. Second stop: I walked to the trailer selling gold-plated horse ID bracelets and I gave them Nicky's show name, Cloud 9, so I could wear a replica of the one I once wore for him. My first best friend. It looked exactly the same as my old bracelet, though this one was shiny and stiff, no mold or softness or scratches to the plate.

Soon after, I visited my mother's house in Long Island and woke up in the morning to see my bracelet aged overnight. Delusional thinking, I thought. Impossible.

But my mother had saved my old bracelet, had it out on my nightstand all along, though I'd never noticed it. Decorative.

I'd swapped them by accident—at some point, I did. Fastened the old bracelet on without even knowing.

I'm not sure which one I wear now.

My wife was a show kid and a groom and a champion and a professional trainer. Yes, I met her through horses; when I came back to them, she was my trainer. She's all movie-version Sonora. All guts. But there is nothing she hates more than the horse show industry.

When my father died, she wrapped my legs in polos because I didn't own boots or half chaps anymore, gave me a leg up on Hank. We rode through the mountains of upstate New York with no numbers strapped to our backs. We just rode. *Everyone knows horses are therapy.*

It was. They are. Sometimes, even for those who don't deserve it, they are.

Sonora did not attribute her peerless diving abilities to other senses strengthening. She attributed her skills to an obsessive attention to detail, to memory. She knew when her husband Al was boiling potatoes because the big pan carried "a deeper voice than the others." She knew the "dull metallic" of the potato masher hitting counter. She even knew when someone was checking the time because of the rustle of a shirt sleeve, the ticking of a clock drawing nearer, then more muffled in the pocket.

"All a person had to say was 'red,'" Sonora wrote, "and immediately the color flashed across my mind; like Mother's cannas along the backyard fence; red like the ribbons we tied on Christmas packages; red like the dress I bought when I was sixteen which was supposed to make me wicked.

"All description is based on comparisons," she continued, "and I had the basis for comparisons."

Real Sonora kept diving those horses with no vision for eleven years, but couldn't eat soup. That was what broke her wild heart, what surprised her most. She couldn't level a soup spoon blind. It was simple like that.

I quit riding horses at thirteen, after an accident. I'd stopped showing and my horses and ponies were sold, gone, but my school had a riding program, so I exercised other rich peoples' horses for them.

One day, after school, I was asked to work a horse named Sprout. The owner told me Sprout had behavioral problems, and I mounted as I always did. I jumped one course and tried for another. Sprout was bad; Sprout fell and I fell and the jump fell, too, and he dragged me by a stirrup; he crushed my ribs in the fall, he bucked like a maniac, but these sentences are in the wrong order; they carry the wrong active subject. He didn't do anything I didn't direct him to do. Sprout was lame, injured; I rode him anyway. He had the wrong bit. I pushed him at the fences. I kicked and spurred and smacked with my crop. I jerked the reins all wrong, and too soon.

Aia ke ola I ka waha; aia ka make i ka waha is another Pukui Hawaiian proverb. It means, "Life is in the mouth; death is in the mouth."

And another: Aia ke ola i ka ihu o ka lio, or, "One's life is where the horse's nose points."

There is one video I found on the internet of Real Sonora and her high-diving horses. Someone has edited the footage so Neil Sedaka sings the ballad "You" over the flickering tape, and it's not one graceful motion the way it looks in the movie. Lightning actually *stops* at the top of the pier, drags her front legs over the edge in painful slow motion. I cannot hear the audience cheer but I know they are cheering, and Lightning resists, looking stunningly terrified. "You showed me how to live again," sings Sedaka. "You gave me strength when I was falling." Sonora in her helmet pushes forward until Lightning not exactly dives, but slides off the ledge and into that pool, neck twisted, hooves digging at air.

Yes, is the true answer. The only answer I can find. The horse does all the work.

HORSE GIRL: AN INQUIRY
CARMEN MARIA MACHADO

As if this big
dangerous animal is also a part of me,
that somewhere inside the delicate
skin of my body, there pumps
an 8-pound female horse heart,
giant with power, heavy with blood.

—"HOW TO TRIUMPH LIKE A GIRL," ADA LIMÓN

I told him I never bit anything but grass, hay, and corn,
and could not think what pleasure Ginger found in it.
"Well, I don't think she does find pleasure," said
Merrylegs; "it is just a bad habit. She says no one was
ever kind to her, and why should she not bite?"

—*BLACK BEAUTY*, ANNA SEWELL

WHAT I WANTED

I wanted, more than anything else, to be a horse girl. I wanted
to ride horses, to own a horse, to go to a stable after school, to be

known as a girl who went to the stable after school. I wanted to outrun a fire or a flood or a bad guy on a horse; I wanted to leap on a horse in order to gallop somewhere to warn people about impending danger. I wanted to groom a horse, to keep her clean and safe and well exercised. I wanted my horse to be distinctly colored, sired under mysterious and auspicious circumstances, birthed under a peculiar star. I wanted to name her, to have people know that name and know my name, too, as the one who had chosen her and been chosen by her. I wanted to know her fears, her bad habits. I wanted to know her from birth, like a sibling or a child. I wanted a horse to feel affection for me; to nicker when I grew near and let no one ride her but me.

WHAT I GOT

For my eleventh birthday, I was gifted a trip to a weeklong horse day camp in New Tripoli, Pennsylvania. Later that summer, my mother drove me every day half an hour out of town. The smell of the farm—fresh green air, the sharp stink of manure—was the most exciting thing I could imagine. It smelled like someone else's life.

WHAT I LEARNED, OR WHAT I REMEMBER OF WHAT I LEARNED

How to gently scoop their fetlocks into our hands and clean their hooves with a pick to prevent thrush. (It had all the satis-

faction of peeling Elmer's Glue off its own ridged, orange tip.)
How to groom them. How to post—that is, to move up and
down with the movement of the horse as they canter. How you
should never walk behind a horse, lest they kick you. How
to put your feet in the stirrups—ankles angled down. How to
steer the horse in both the English and Western styles. How
to put on a bridle and saddle and remove them again.

WHAT WAS MINE

A water bottle and packed lunch I brought with me every day.
The clothes I wore. A bike helmet.

WHAT WAS NOT MINE

Every camper was assigned their own horse for the duration
of the camp. Mine was a bay—reddish brown, tipped with
black. His name was JC, but his show name was One Sexy
Thang.

I didn't know before this camp that horses had barn
names and show names. It tapped into some pleasurable core
I didn't know existed—the kind that finds drag identities and
Roller Derby handles so magical now. To have one name you
are called in your everyday life, and then some second, better name—a fragment of a phrase, a lovely idiom or a turn
on one—it seemed like some new dimension that horses possessed. Two names, just like that.

It should be said that I didn't fully understand his show name. I understood, very roughly, what sex was, but I had never heard the adjectival form. When I asked the young man helping us—in my memory the son of the owner of the farm—what "sexy" meant, he said it meant that you looked nice. Later that afternoon, while the young man was explaining horse anatomy, JC dropped his penis—surprising, massive—and peed in gushing, endless waves as the young man's voice rose louder and louder over the sound.

SHOW NAMES

Talk of the Town. All That Jazz. Shadow of a Doubt. Ask No More. Chief Lord and Sugar. Return to Sender. A Cut Above. Blood Diamond. Mark My Words. A Fine Turn. Maybe Tomorrow. No Autographs Please. Nine of Cups. North by Northwest. Note to Self. On Second Thought. Be My Daddy. Between the Lines. Runs with Scissors. Change the Song. Satisfaction Guaranteed. Coffee and Cigarettes. Since You've Been Gone. The Velveteen Rabbit. The Way We Were. To Catch a Demon. Friday Night Lights. King's Mistress. Look Who's Talking. Lost in Translation. Made You Look.

WHAT WOULD NEVER BE MINE

At the end of camp, when my parents came to see me ride and show off what I'd learned, I told my father that JC was for sale.

Could we buy him, please? My father looked at me like he'd never seen me before in his life. No, he said. Please? I asked. Carmen, he said in a measured, patient voice, where would we put a horse? It was a very good question—an engineer's question. It did not try to shame, or laugh, or encompass the ridiculousness of my request; rather, it was specifically designed to root out a fatal flaw without having to waste time pulling apart the very foundation of the query. I left that day without a horse, and have been horseless ever since.

PHOTOS OF ME ON OR NEAR HORSES

From the New Tripoli horse camp, there are three images that remain in my possession. They were all taken on the same day, as I am wearing the same outfit in all of them: jeans too short for my legs, white crew socks, white sneakers, a pink shirt that says "Yes I Can!" which was the name of the Girl Scouts cookie initiative in 1993, and a knock-off Tamagotchi—an ambiguously specied dinosaur—on a watch. All the photos were almost certainly taken by my mother, and most likely on the last day of horse camp.

In one photo, I am riding JC in an arena with a helmet on my head and a riding crop in my hand. The camera flash has caught his *tapetum lucidum*; his eye is white and sharp as a tooth.

In another photo, JC is sticking his head out of an opening in his stall and is very close to nibbling my ear. My baby sister is standing next to me. The photo is not in focus.

In another photo, this one in focus, I am performatively grooming JC with a round brush. I am staring directly into the lens. I am clearly wearing lip gloss and tiny gold hoops in my ears. Again his eye glows; this time, a silver crescent moon.

THE FIRST LESSON

One day at the horse camp, I wore a pair of overalls with a tank top. It made me feel very rugged and cute. (My mother had refused to buy me the "riding gear" that I begged for, and this was the next best thing.) That afternoon, the young man from the first day told me, an eleven-year-old-girl, that I looked sexy. After he walked away, I remember feeling feverish and uncomfortable; the way I'd felt the last time I'd been laid up with a stomach flu, naked underneath a woven afghan, a tiny, sour-smelling wastebasket on the floor near my head.

He didn't do anything else, though, after that. This isn't that kind of essay.

A NASTY THING

At the camp, I hesitated to put the bit in JC's mouth, because I'd read a glossy, faux leather–bound edition of *Black Beauty*, which went on about the foul nature of bits *at great length* and *in the horse's voice*. ("[A] nasty thing! Those who have never had a bit in their mouths cannot think how bad it feels—a great piece of cold, hard steel as thick as a man's finger to be pushed into one's mouth . . . no way in the world can you get rid of the nasty hard thing. It is very bad! yes, very bad!") The farmer's son insisted that the horse did not mind, he was used to it, and I wondered what it meant to be used to something unpleasant; what it meant to know something was awful but then let that knowledge go down, and under.

MORE PHOTOS OF ME ON OR NEAR HORSES

In another set of photos, I am at my grandfather's farm in Mineral Point, Wisconsin. By then he had sold the farm to a man who trained horses for the Barnum & Bailey Circus, though my mother and her eight siblings and their children were permitted to visit, which we did every summer. The fact that my mother grew up on a farm always amazed me. But they didn't have horses, my mother said, though they did once have a pony named Strawberry Roan. I don't remember if they rode her, or what happened to her.

In one photo, I am sitting on a horse with my kid sister in the doorway of a barn. I am tan and in overalls. My sister is missing a tooth. My arms are wrapped around her, clutching the front of the saddle. The horse's ears are back in faint alarm. In the background is one of my mother's older brothers, his eyes closed in a forever blink.

In another photo, a long herd of distant horses—speckled, blazed, starred, striped, bald-faced—bisect a pale green meadow. I almost certainly took this photo myself, with a disposable camera.

In another photo, I am standing shyly in front of two massive Clydesdales as if they are my bridegrooms and I can hardly believe my luck. I am leggy as a colt. My long hair is up, I am wearing large glasses, and a halo of frizz floats around my face. It was almost certainly shot with the same disposable camera, though I—obviously—did not take it.

THE SECOND LESSON

The first time I ever wore a tampon, at age twelve, I walked around the house bowlegged, agitated from the sensation. My mother said I looked like I'd been in the saddle all night, which implied a kind of agency and rugged self-sufficiency that seemed as distant as adulthood. Cowboys rode in the saddle all night, if they had to or wanted to, but they did not have to wear internal menstrual products in order to participate in swim class. ("You can't just bleed everywhere," my mother said.) I had been given a body. Choices had been made for me.

A CONFESSION

I am exasperated that my essay about girls and horses contains inappropriate compliments from older men, sexual innuendo, sexual exploration, menstruation, outfits.

SEXUAL EXPLORATION?

I'm getting there.

WHAT OTHERS HAD

My best friend, Fiona, lived an enviable life as the only child of two older parents. My own parents tried to temper my envy—it must be so lonely, just her, they said, and her parents are older and will die much sooner than we will, plus they smoke cigarettes. I didn't like the stale and ashy smell of her house, true, but I did love her shelves full of collectible Breyer horse figurines (!), her regular horseback-riding lessons (!!), and a massive mural of a white unicorn on her bedroom wall (!!!). It would, I thought, be worth it to be lonely, and to have my parents die sooner than other people's parents, to have those things; after a while, the cigarette smoke would almost certainly become barely noticeable.

WHAT A HORSE GIRL IS

Lithe teenage girls with impossible French braids, who as adult women will tell men they're not like other girls. If you were to lean close and breathe deep, she would smell like heterosexuality, independence, whiteness, femininity. Then a heart note of old-school feminism, both admirable and dated. And at its base, the warm whiff of rural wealth, the kind that thinks of itself as different from other kinds of wealth even though it is very much the same.

WHAT A HORSE GIRL IS NOT

Fat. (Though I wasn't, not really, not yet.) Queer. (I didn't know I was, but maybe they knew?) Latinx. (Despite a robust nonwhite tradition of horse riding that dates back centuries—vaqueros, rancheros, charros, Black cowboys, Indigenous horse riders, paniolos from Hawai'i—we think of horseback riding as being necessarily the sport of the wealthy—and, by extension, whiteness.) Me, though I didn't know it yet.

WHAT A HORSE GIRL IS NOT, ALSO

When my best friend from college got married in Shenandoah National Park, I took my then-girlfriend with me. By then I was openly queer, very fat. One of the offered activities was horseback riding, which I insisted on doing. "I used to love it," I said to anyone who listened. "I loved horses so much when I was a kid."

My girlfriend was amazed that I wanted to get on a horse and ride it down a trail. She'd never ridden a horse before. That's some gentile shit, she said. I was such a shiksa, she said. She was always calling me that; she'd read too much Philip Roth. We rode on a trail with a guide, each on our own horse, and I was amazed at how much I remembered from camp, all those years ago. Posting up and down on the horse's back; the sensation that I was this tall, magnificent creature, that all of her blood and bone and muscle were my blood and bone and muscle. My horse seemed a little old and a little tired, but it didn't matter; I was a little old and a little tired, and anyway, I felt pleasure down to my marrow.

Years later, when I was dating another Jewish person, I told them about my ex calling me a shiksa—her voice teasing, how she'd occasionally follow it with the word *goddess*, the way she insisted it was a compliment because I was so beautiful. They listened with a furrowed brow and then told me that it was not, precisely, a nice thing to call someone. The phrase did not convey the tender, loving teasing I'd hoped it had. They seemed a little sad about it. More recently, someone else told me that shiksa also implies whiteness (and blondness, and a kind of WASPy sensibility), and I found myself tallying points about my body and my family and my identity and my upbringing and the things I loved in mental columns of whiteness and non-whiteness. Later, someone else disagreed that shiksa implied whiteness, though they acknowledged it was an unpleasant word, and I scribbled a messy line all over the math. All I understand is that there is so much that I will never understand.

So instead of parsing the memory of that afternoon in the mountains along lines of race or identity or beauty, perhaps this is the only way to think about it: riding behind my then-girlfriend on the only creature that could hurt me worse than she could.

HORSES I WANTED

I remember, as clearly as anything, Fiona's horse-themed birthday party, in which a small, vinyl horse figurine was placed at the head of each of the paper plates that ringed the table, where a wine glass might go. I rushed to sit at the seat with a jet-black one, but Fiona's mother gently guided me to a gold-and-white one instead. As an adult I recognize that a palomino is a perfectly fine and even beautiful horse figurine to possess—writing this I am possessed by a powerful urge to have it, still—but back then I wanted the black horse because it was clearly the superior creature. Everyone knew that black horses were the most elegant and rebellious, the fastest and most brave, the kind of horses that entered into legend. Fiona sat at the coveted seat. He was hers, now. Later, when we were playing some game, I picked up the black horse and touched it to my lips, like a rosary.

METAPHORS THAT DON'T FIT ANYWHERE ELSE

Fiona was the person who first told me about sex. She told me that it meant getting naked and kissing. More than once we played "mom" and "dad," lying on the floor in my basement or hers and lifting up our shirts for each other. It is the first sexual memory I have.

I haven't talked to her in twenty years. I do not know if her parents are dead or alive. I do not know if she still rides horses or loves horses or has her Breyer figurines. I do not know if she remembers me and the keenness of my envy.

HORSES I HAD

The aforementioned vinyl palomino, a different scale than any of the horses that follow, and which I refused to give a name.

A set of hollow plastic horses that came with a Western stagecoach, which I later repurposed for a fourth-grade history project, in a diorama about the Amish.

A rideable, child-size horse Flyer on springs and a metal frame, upon which my siblings and I would launch ourselves day in and day out, the entire thing squealing in protest as we rode it wild-eyed, scrambling to get on first, hauling each other off as often as possible.

A set of Breyer knockoffs my parents gave me one Christmas: a chocolate-and-white dappled Appaloosa father frozen in a high step; a bright palomino mother with her head dipped, demure; a gawky baby with the mother's coloring and the father's spots. All had soft, lifelike hair that could be brushed, braided, and managed, which I did with great care.

A plastic, electronic horse of unknown provenance—brandless, likely older than I was—that was too big for a Barbie and too small for an American Girl doll, with a cord and controller that when activated made her legs move stiffly and eventually caused her to fall over with a tremendous crash, which I bought from a yard sale with five dollars of my own money.

A unicorn snow globe that was surrounded by ceramic

roses that my dad got me to apologize for going away on a business trip for my birthday. We bought it at a shop that exclusively sold fancy snow globes. He tried to persuade me to buy a pegasus that played "Wind Beneath My Wings," but I insisted on the unicorn, which played "My Favorite Things."

A My Little Pony: Blue Belle, with her plasticky, cupcake-scented haunches. When her scent faded I tried to refresh it with my mother's perfume. She caught me and took it from me, dangling the toy from her pinched fingers like Achilles's mother dipping him in the River Styx. I had ruined it, she said, and threw it away.

A series of images advertising vaguely racist Native American–themed ceramic horse figurines and collectible plates for sale in *Parade* magazine, which I clipped out and kept in a photo album.

HORSES I MADE

I had a small library of horse-drawing guides. I was very partial to horse heads and necks; the architecture of their jaw, the swooping parabola of their necks. Decapitated horse heads, à la *The Godfather*, littered my grade-school notebooks.

HORSES I WAS

Later on, after Fiona, my friend Margaret and I often played a game in which, on land, we were alternately girls with horses or unicorns, or horses or unicorns themselves, which translated to dolphins in the pool where we swam all summer. I often called my horse, or horse-self, or dolphin, or dolphin-self, Fire Maiden, which to me seemed the most eloquent thing you could be named, or be.

ARE UNICORNS HORSES?

Unicorns are horses that can only be ridden by virgins.

ARE HORSES UNICORNS?

Horses don't care what you've done, or what's been done to you.

WHAT ARE HORSES?

A species of odd-toed, ungulate mammal, primarily domesticated, belonging to the taxonomic family *Equidae*. Useful, expensive, dangerous. Beautiful.

WHAT ARE HORSES TO ME?

An admittedly impressive animal. Something that held value for me as a girl that holds less value for me now, as a woman.

WHAT WERE HORSES TO ME?

A creature that I believed could hold me up, make me better, give me context.

A desirable thing that, had I become part of them as I wanted to (through knowledge, ownership, skill, love) with all of my tiny, terrible heart, could have permitted me a streamlined relationship with my own gender, and offered me a kind of highly specific authority that would have softened the edges of a strange and unpleasant adolescence.

A metaphor. Or, multiple metaphors. The way the desire for independence and power and control can become tangled up in domesticity, into the things that can weigh you down. How means of movement need to be bought, fed, cared for. How you can easily acquire a false sense of control, how appealing it is to let something into your life that could crush you if it wanted to, or even if it didn't.

BUT WHAT WERE THEY, REALLY?

"As one woman I interviewed said about her beloved childhood horse, 'He was my beauty'—not in the sense that he was her beautiful possession, but in the sense that his beauty became hers too."[1] A beauty that could be borrowed.

---

1 *Horse Crazy: Girls and the Lives of Horses*, Jean O'Malley Halley

CAN YOU BORROW BEAUTY? IS BEAUTY A THING THAT YOU CAN ACHIEVE THROUGH DESIRE, POSSESSION, OSMOSIS? CAN YOU BE BEAUTIFUL WITHOUT A BEAUTIFUL CREATURE BENEATH YOU?

I'm still not sure I know.

BREYERFEST OR BUST

LAURA MAYLENE WALTER

To travel to the Kentucky Horse Park on the outskirts of Lexington is to drive through acre upon acre of rolling pastureland. This is the heart of Kentucky's horse country, a land of luscious bluegrass, rich brown fencing, stone stables, and priceless Thoroughbreds. The Kentucky Horse Park itself is a sprawling complex of arenas, stables, shops, museums, memorials, and training facilities. The park hosts equestrian demonstrations of all disciplines, prestigious competitions, and other major events like the one that brought me there for the first time in 2019—BreyerFest, a three-day celebration of horses both real and plastic.

I was drawn to BreyerFest not only because it sounded like a kitschy getaway, but because it presented the rare chance to be a tourist in my own past. I was in my late thirties, horses were no longer an integral part of my life, and my childhood collection of Breyer models was relegated to a single box in the attic, but even so, I was reluctant to let go of my former horse girl self. If revisiting that part of my past meant attending a festival where attendees worked themselves into a frenzy over plastic horses, then so be it.

Which is how, only moments after arriving on the grounds, I found myself staring transfixed at a model horse diorama competition. The dioramas were set up on folding tables that twisted through the park's visitors' center, an impressive display of the creative possibilities unlocked by dressing up toy-size plastic horses. One model horse was styled to look like Lady Gaga, another like Elton John, and yet another as an astronaut with a little space suit and bubble helmet. Elsewhere, equine firefighters worked to save model horses from a burning building, an act of heroism that included a fire truck and ladder, a life net to catch jumpers, and a paramedic foal using its mouth to pull an emergency blanket across another horse's back. My favorite entry depicted none other than the diorama contest itself, in which a plastic donkey wearing a judge's badge and diamond earrings scrutinized a collection of mini dioramas.

My husband and I happened upon this display after making the long walk from the parking lot, where our hatchback was dwarfed by minivans and SUVs with messages like "BreyerFest Bound" painted on their rear windshields. As festival first-timers, we thought we'd get oriented at the visitors' center. Instead, we found ourselves gazing at a model horse spray-painted gold in a cardboard celestial backdrop filled with cotton clouds, twinkle lights, and dangling planets. "My diorama is out of this world," the entrant had written on the accompanying card, and I couldn't exactly disagree.

This year was BreyerFest's thirtieth anniversary, and over the course of three days, upward of 30,000 people would swarm the Kentucky Horse Park to purchase limited-edition models, watch riding demonstrations, attend model repair and painting workshops, meet the real-life horses who inspired some of the models, and commune with other collectors.

Peter and I left the visitors' center to cross under a Breyer-branded welcome arch and join the crowds outside. I paused to take a photo of a life-size inflatable horse soaring over an inflatable jump. All around me, people wore shirts emblazoned with horse heads and carried bulky bags full of new Breyer models. I walked on, my attire horseless, my hands empty, and felt unmoored. No matter how much horses had once been a part of my identity, I had become an outsider, a stranger in the land of plastic ponies.

For the uninitiated, Breyer horses are ubiquitous plastic toys that have delighted both children and adults for the better part of a century. The first model, a palomino in Western tack, was introduced in 1950 as part of a mantel clock sold at Woolworth's; it proved so popular that Breyer began manufacturing stand-alone plastic horses. Today, thousands of Breyer models are available in nearly every breed, color, and pose imaginable.

Before hitting the road to Kentucky, I tried to explain the allure of Breyers to friends who could not grasp how plastic horses support a three-day festival. The models are realistic, I said. The molds are sculpted by artists and scaled accurately for each breed. They are not mere toys but are incredibly detailed, with visible muscle tone, bodies posed to suggest movement, and the illusion of life emanating from their painted eyes. Accessories like miniature halters, blankets, and tack were also available. You could purchase a scaled-to-size stable with miniature feed buckets and bales of hay, or even tiny leg wraps to protect your model horse before loading it into a plastic trailer. The Breyer universe was a stage for the imagination, providing a physical outlet for the dreams so many horse-obsessed girls share.

I had once been one of those girls. As a child, I received a new Breyer horse as a gift for every birthday and Christmas, and I saved to buy more whenever possible. Back then, I could obtain a Traditional model for $9.99, so each crumpled ten-dollar bill that landed in my hands transformed, in my mind, into a new model horse. I amassed three dozen Breyers over the years, including a mix of the Traditional 1:9 scale models along with some smaller versions. I could still recall my favorites: the Clydesdale stallion, the rearing Appaloosa, Misty of Chincoteague, the prancing Arabian mare.

When BreyerFest was founded in 1990, I was nine years old, a horse-crazed girl growing up in Lancaster, Pennsylvania. Had I known BreyerFest existed back then, I surely would have clamored to go, and had that happened, I would have fit right in. I, too, would have marveled at the live horse demonstrations, geeked out over my favorite models, and begged with fiery passion for that year's commemorative model—or for any new Breyer horse, that magical toy molded from girlish dreams and cellulose acetate.

It wasn't until decades later that I'd convince my husband to give up a summer weekend to drive the five hours from our home in Cleveland to Lexington, where we found ourselves wandering the BreyerFest grounds as an entire family of five—mother and father included—sashayed past with faux horse tails attached to the backs of their pants.

Many of the flashiest sights at BreyerFest took place in the covered arena. The arena's perimeter walls were plastered with BreyerFest ads while a giant banner at the back showed a photo collage of girls hugging both real and model horses. A packed schedule of breed and riding demonstrations filled the arena throughout the weekend.

The attractions were eclectic, to say the least. Peter and I stepped inside the first time to see a Morgan horse standing on a platform next to a glowing American flag while "I Will Wait" by Mumford & Sons filled the arena. Later, a miniature horse named Hughie donned a silver cape and leapt over fences; Oliver, a mounted police horse, navigated an obstacle course; and trick ponies galloped the length of the arena as their riders performed acrobatics.

During a Western dressage demonstration, I leaned over to Peter and whispered that riding dressage in Western tack struck me as a bit ridiculous.

"Maybe that's snobby of me," I said, "but I can't help it."

He nodded. "I understand because—and maybe you won't like to hear this—to me, *everything* people do with horses seems ridiculous."

I laughed, but then I tried to view it all through his eyes. Attaching a cart to a miniature horse so it could pull you around in circles? Ridiculous. Donning shiny black boots and a formal jacket to guide a horse over a series of low jumps while being judged on form? Ridiculous. Dressing your Arabian in filmy scarves and tassels as part of a "Native Costume" competition? Definitely ridiculous, and served with a heaping side of cultural appropriation, too.

That Saturday, the festivities also included a live performance by Nashville country music artist Templeton Thompson. She strode into the arena in jeans and cowboy boots to serenade Cobra, a dark bay gelding with flowers braided into his mane and gold glitter splashed across his rump. Cobra was famous for his transformation from wild mustang to dressage champion, and BreyerFest 2019 marked his official retirement from competition. As Templeton crooned into the mic, Cobra's rider guided him through some dressage movements, then cantered straight down the side of the arena to drum up applause from the crowd.

Later, Cobra was adorned with a wreath of flowers as part of his retirement ceremony while Templeton sang her original song "Girls and Horses," a tribute to the unbreakable girl/horse bond that was on display throughout BreyerFest. Everywhere I looked, in fact, I saw girls—girls who stood by horses like faithful handmaidens and girls who clutched Breyer models to their chests with the fervor of the religiously possessed.

I knew exactly how those girls felt and how much horse love brimmed inside them. I'd practically grown up at stables thanks to my mother, who'd loved horses all her life but was an adult before she began riding. Once she started, she allowed that pent-up horse love to consume much of her daily life. She bought Amrieh, a chestnut Arabian mare, when I was four years old. Within a few years she bred Amrieh, named the filly Shilo, and spent the rest of her life doting on both horses with the whole of her heart.

Together, my mother and I passed many hours driving to and from stables. Together, we walked the length of pastures to track down our mares, we groomed and tacked up in barn aisles, we circled riding rings or headed out on the trail. Gentle, reliable Amrieh became my horse while my mother rode Shilo, making two sets of mothers and daughters between us.

That my mother and I were able to share so much time together around horses now strikes me as a kind of miracle. I was twenty years old and in the middle of my sophomore year of college when she died, unexpectedly, of cancer. Keeping both our horses at that point was impossible. I gave Shilo to my equestrian cousin, and I paid for Amrieh's retirement board from my portion when my brothers and I sold our childhood home. Amrieh was in her late twenties by then, and she only lasted about a year before she fell ill and had to be put down. And just like that, I entered adulthood with neither my mother nor my beloved horse, both of whom had taught me what it was to love and be loved unconditionally.

By 2019, my memories of those carefree days spent in the company of my mother and our horses seemed unreal. But there at BreyerFest, as I watched so many girls wander the park grounds with their own mothers, the distance between my present and my horse girl history seemed to be narrowing. I could sense the ghost of my past self everywhere, from the

thunder of hooves in the arena to those model horses forever
frozen midstride.

The BreyerFest arena also contained the marketplace, a con-
gested space full of vendors hawking all manner of models
and horse-themed clothing and gifts. I lingered at a stand selling custom models, Breyers that had been hand-painted by
the seller. I considered shelling out $45 for a mare painted a
lovely shade of purple with a silver mane and tail. It was a
color scheme I would have gone wild for as a girl.

Just before reaching for my wallet, I paused to examine
the model more carefully. Up close, the paint job was sloppy,
especially around the eyes, where dabs of silver made the
mare look like she'd been blinded in some kind of molten
metal accident. Perhaps it wasn't something I'd have noticed
as a child, but my adult self was disappointed. With a touch of
resignation, I put the mare back on the shelf. Maybe I'd have
better luck in the official store—the shop younger festival
attendees nicknamed the Ninja Pit of Death, presumably for
the fierce competition that arises when thousands of collectors
battle it out for limited-edition models.

Peter, who was exhibiting astonishing patience and good
humor as I dragged him around the park grounds, agreed to
venture with me into the Ninja Pit of Death. While he was not
a horse person himself, he'd always supported my affection
for horses, and he indulged me on the rare occasion I accessed
the vast stores of horse care and riding information that had
taken root in my brain decades ago. That knowledge had little
reason to surface in my adult life as a writer and editor, but

when it did, it felt like opening up a part of my past to Peter, a way to show him where I'd come from and what had once mattered most to me.

To gain access to the hallowed Ninja Pit of Death, we stood in a holding pen across from the arena. It felt like waiting in line for a roller coaster at an amusement park: we were corralled in a maze of gates while giant fans blasted overhead to provide some relief from the heat. While we waited, I eyed the middle-aged woman behind me. We'll call her Barbara. She wore the knowing air of a festival regular as ostentatiously as her outfit, a shirt screen printed with a horse head, and a fanny pack clipped around her waist.

"This is my first time visiting the official store," I told Barbara as the line inched forward. "I hear it can get intense."

Barbara nodded. "Some of the special-run models are limited to one purchase per person," she explained. "But if you're serious, you can buy one, check your purchase, and get back in line to do it all over again."

Barbara, I learned, was definitely serious—she'd been attending BreyerFest for years, where she bought models to resell online for a profit. She owned hundreds of Breyers, which made me think with some pity of my neglected collection at home. I hadn't even stored the models properly; I just tossed them in a box and that was that.

"Some people leave their bags there before getting in line again," Barbara continued, nodding to the bag-check booth next to the holding line. "But I don't trust it. I walk back to my car every time I buy something and leave my models there instead."

I estimated the parking lot was nearly a mile away, which meant Barbara was getting her steps in. Clearly, I had found the dedicated BreyerFest expert who could answer my questions.

I flipped open my official BreyerFest program and pointed to the column labeled "Autographs/Meet & Greet at Stables," which provided a schedule of the real horses one could meet to obtain an autograph on the Breyer model that was made in the horse's likeness.

"How can a horse autograph a model of itself?" I asked in all seriousness.

Barbara looked at me as if realizing I was perhaps not very bright. "The horse's trainer or owner signs it. Not the horse."

"Ah," I said, trying to hide my embarrassment. "That makes sense."

I was about to ask if the trainer/owner signed their own name or the name of the horse, but Barbara had already embarked on another story, this time about an owner of a real-life Breyer horse she'd met at the festival some years back. There'd been some mix-up that had delayed the owner from receiving the brand-new Breyer model based on her horse.

"Can you imagine?" she asked. "To not even have the model of your own horse!" Fortunately, Barbara went on, she had braved the official store earlier that day and bought two of the models in question. She gave one to the owner, who returned the favor by autographing the other model on the sly and outside the official channels, which was apparently pretty scandalous in Breyer world.

Around this point, I glanced at Peter, who appeared to be staring straight into the sun with a vacant, I-guess-this-is-my-life-now expression. The line wranglers chose that moment to release our group from the pen, finally giving us access to the Ninja Pit of Death.

"Good luck," I told Barbara, and then we walked up a ramp behind the arena in an orderly fashion, which seemed anti-climactic. True, we were visiting the store after the initial rush, but where was the desperate running, the shoving, the hair pulling? Instead we were a bunch of overheated white people trying to find a bit of shade before finally making it inside a wonderland that turned out to be . . . another store filled with model horses.

Peter and I browsed, passing editions marked LIMIT ONE PER PERSON. I bought a few toy ponies as gifts for my nieces, but I didn't purchase any "real" models, meaning the 1:9 scale horses packaged in yellow boxes with clear plastic fronts—those delicious, full-of-promise boxes that had been the center of my materialistic desires as a child. I was willing to drop the $65 or so on the right model, but none pulled at me strongly enough to make the call. I was no longer a child for whom a plastic toy could come to life, and this made me feel as though a light inside me had been extinguished.

There was only one way to see if at least a spark of that light remained: head to the stables and check out the real-life Breyer horses.

The smell of horses, of dust and hay and even manure—it fills me with a melancholic nostalgia each time I encounter it, and BreyerFest was no exception.

At the stables, the horses who'd inspired models wore

special blue-and-yellow halters that read "Breyer." As I admired King, a palomino pinto and one of the stars of the Trixie Chicks Trick Riders, I had to admit that the halter made him look like he'd stepped to life right out of a Breyer box. It was a glimpse of what it once felt like to gaze at a model horse and believe it had a spirit nestled within its molded plastic shell. In a flash I was a girl again, saving my chore money to head to my town's eclectic general store to select a new Breyer model. I could still feel the tight twist of plastic ties that secure the model's legs inside the packaging. I could still recall the sensation of my hands sliding over the smooth body that felt light and alive and ready to play.

"There is something inexpressibly charming about a plastic horse," collector Nancy Atkinson Young wrote in the preface for *Breyer Molds & Models: Horses, Riders, & Animals, 1950–1997*, a guide published in the 1990s. "Indeed the first thing many a Breyer collector does upon acquiring a new model is to plunge her nose into the box and take deep whiffs of that mystical perfume of cellulose acetate."

I know that smell as well. It makes me think of tack shops, saddle soap, and the velvety inner knee patch on a pair of breeches. It's the scent of plastic, but it also conjures the smells of leather, and dirt, and the dust that rises from a horse's thick winter coat during a good currying. The memories I associate with that smell take me right back to those days spent at the stables with my mother. In the summertime, we coughed through clouds of fly spray. In the cold dark of winter, we made the horses warm bran mash drizzled with molasses as a treat. If one of our mares colicked, we'd stay up until dawn taking turns walking her up and down the aisle until the vet arrived. We also mucked out stalls to reduce the board payments, endured family tensions over money, and switched barns multiple times in search of a more affordable price—but overall we led a charmed, horse-filled existence.

When I wasn't at the stable, I had my Breyers, horses that could never colic or buck or bite, not to mention rack up expensive board and vet bills. With a model horse, I could imagine myself as a perfect rider, one who never fell or had bad form or grew frustrated. In my mind, I gave perfect cues, my horse was always balanced and on the bit, and I was never, ever afraid.

I continued to pursue this unattainable perfection as an adult during the sporadic periods I took riding lessons or half-leased a horse. But no matter how many times I tried to return to riding, I never recaptured the sense of freedom and joy I'd experienced as a child when I rode with my mother. Instead I was often self-conscious, sometimes even fearful, and I struggled to meet the financial and time commitments that horses demand. When I had a job that supported the cost of riding, I didn't have enough time to spend at the stable; when I had the time, I didn't have the money.

The girls all around me at BreyerFest, meanwhile, were still in the full flush of their horse obsession. They were everywhere, wandering the festival while carrying their model horses like talismans. The sight was so ubiquitous that Peter and I returned to the marketplace to buy tiny $2 unicorn models so we, too, had something to hold on to. Laughing, we clutched our unicorns as we walked the park grounds, thus joining the ranks of the Breyer girls and their beloved plastic companions.

We laughed, but it wasn't entirely a joke to me. The way these girls held and admired and gazed lovingly at their new models felt familiar. I understood how an injection-molded plastic toy could conjure complex fantasies surrounding horses—to achieve flight on the back of a galloping horse, and to become strong, beautiful, and powerful by extension.

I understood those girls because I had been the same, and in fact that child from the past was still alive inside of me. She's the one who prompted me to keep that boxful of Breyers for decades. She's the one privileged enough to not only have grown up around horses but to have had a mother who shared and facilitated a deep love for them. And she's the one who decided to journey across state lines on a hot July weekend to enter a world where horses are worshipped like celebrities or gods.

The day carried on, the sun boiling. While Peter waited in line at a food stand, I wandered over to some shade. Only after plunking down on the ground did I notice Cobra, the famous mustang, waiting nearby with his handlers. I remembered how the announcer had called Cobra "a true American Cinderella story" and said that he'd risen "from rags to riches, from wild to world famous." Before he became a star, Cobra was a wild mustang held by the Bureau of Land Management, which administers federal adoption programs for wild horses and burros. After being passed over for adoption three times, Cobra was labeled "unadoptable" and left to languish in a government corral—all until the right trainer took a chance on him. Now Cobra was a champion, an event headliner, and had his own Breyer model to boot. And he was right next to me in all his glorious bay flesh, enjoying a quiet moment away from his adoring fans.

I stayed off to the side to signal that I wasn't there to harass Cobra for an off-the-books autograph or photo op. I was a woman, sitting alone, reflecting on the day and the past. So much about BreyerFest struck me as absurd, foreign, and far removed from my adult life that centered on bookish pursuits like writing, editing, and teaching. I laughed at the strange sights and told myself I'd outgrown this. I was better than these adult BreyerFest attendees who were so unabashedly horse obsessed, even still, that they dressed in horse-themed clothing and fought over rare models and spent thousands of dollars every year at a festival devoted to plastic horses. I believed I was different now. But I was not. The smell of the stables, the sight of a familiar blue-and-yellow logo, the model horses capable of coming alive with the merest hint of imagination: this all belonged to me. This was my history. This was who I was.

So I sat quietly and watched girls walk by holding model horses to their chests. Clinging, clinging with unabashed devotion to all that they loved more than anything else in the world.

HUNGRY AND CAREFREE
ALEX MARZANO-LESNEVICH

How I hated the jodhpurs. If you've worn them, you know why, can probably conjure their feel against your legs: the heavy yet too stretchy material; that color like sand or Play-Doh or a pallid doll's hard plastic flesh, a color that has never looked good on anyone, only determinedly inoffensive, the most offensive thing about it. And then the horrible cling of those pants, the way there was just your *shape* in them, the bud bloom of your body right there for you and your mom and the other horse girls to see. Worse yet was that the pants somehow conveyed in their very fabric the suggestion of how they were supposed to look: smooth and elegant, like you were tall and thin and had hair as straight and plain as the pants. Like you were one of the cool blond saplings of a Ralph Lauren ad, part of an idea of America that had taken British tradition and Western adventure and forged a wholesome new myth from it.

Like, in other words, you had money. Money was what the tall black riding boots cost, as well as the fourteen-karat horseshoes that hung on chains from the horse girls' necks and dangled from their charm bracelets. Money meant that your hair was honeyed and your clothes cut well and you'd learned to dance at the country club on the hill, where your father played golf, and that the pants and the helmet and the riding crop and even the horse weren't borrowed, but your very own. The one time I remember going to the tack shop a town over in northern New Jersey—ours striving to be a tony town of cash, always striving, but the other far tonier, my sisters and I raised to be attuned to that difference—I gaped at the perfection of what was on display: tiny plush horses with their tinier, bristly currycombs; the gorgeous gleaming leather of the saddles. Imagine owning your own saddle! Here, thumbing through racks of jodhpurs for ever-smaller sizes, were the girls who never got messy even as they mucked stalls, the girls whose hair stayed neat in their ponytails as they rode the backs of hulking, sweating creatures. Who were long-limbed and loose-limbed and who went to, oh god, summer camp. A tribe of girls, a flock of girls, a stable—the collective noun—of horse girls.

Then there was me.

I looked like a perfectly ordinary girl, if you want to know the truth. I tell you that to remind myself of it, too, so hard it remains to look at photographs from that time. The girl in them has long curly hair that pokes up around her face with frizz and snarls in the back into a dark, knotted cloud. Her glasses are big and round and always sliding down her nose. That nose is freckled, and while it might not have anything as obvious as newsprint on it, it somehow nonetheless carries the suggestion that it belongs in a book. If you could coax her to speak, her voice would barely be a whisper. She prefers the shadows at the sides of the stalls. She seems to want to be swallowed up by the shadows, the way her dark green T-shirt swallows her up before the tight betrayal of the pants.

A borrowed riding crop in her hand, a borrowed helmet on her head, all these rented props that tie her to this role. That riding helmet was hard plastic covered in the soft fuzz of black velvet, an adult men's large, size $7\,^5/_8$, and I still know that size because it always occasioned comment at the store, that a girl should have a men's-size head.

But let's go back to those pants. Because only writing this do I realize I've conflated two pictures, two outfits: the pale gray of Little League stirrup pants, paired with the dark green shirt. Those stirrups were stretchy, they clung, but still, I felt only pride wearing them, never shame. The other kids on the team were all boys, and I took my place with them as though there I might belong, as though the pants fit me the same way they did them, proud to be a pitcher, proud to throw my one no-hitter and collect the signed game ball after, covered in my classmates' scrawls. Why was I happier then, when I looked in the mirror after games, my hair lank and sweaty under the ball cap, my shirt come untucked, dirt patches at my knees, the glove I oiled and kept under my mattress still clenched in my fist? What power did these symbols have, this context, that spun a story that let me forget my body and my unease? And when a retrograde state law swept through our ballparks, banning young women from Little League out of a supposed fear for our developing shoulders, and redirecting us to softball, why did I never take to softball like the other girls, to what was supposed to be the sport for growing girl bodies?

A perfectly ordinary girl. Sometimes when I reach back into the past, when I—let's go for the language, why not—try to muck out the past, this is what I come up against: I didn't understand how I felt back then. I didn't have words for what I felt, or a narrative that could tell me who I was. But now that I do, the signs of what was wrong are everywhere, obstacles laid out to be jumped.

Jo Monahan was sixteen years old in 1866 or so—the historical record is unclear on the year—when he packed up what little he owned, slipped out of a clapboard house in Buffalo, and hopped a train bound west to Idaho's Owyhee Mountains. Nothing exists in his words about his journey, so we will have to use our imaginations to tumble back across time, to tumble into his skin. Picture the clothing he would have chosen: not, of course, the white lace dress the likeness of which ran in the newspaper decades later when he died, the girl in that image Jo's age and with his same face, but hands folded demurely in front of her, the sweetheart neckline of her dress displaying her delicate clavicle, her long blond hair tied softly back. A delicacy that has never been Jo's by birthright, only foisted on him. Give Jo, instead, some rough-hewn Levi's to pull over his legs, a bandana to knot at his neck, a wide-brimmed hat to tuck that hair under. In a few days he'll shear off his locks, transformation begun, but here he must still hide his intent— must still hide himself—from his foster mother's notice. The foster mother, he will protect from this knowledge, at least for now. The foster mother, he loves. It was she who took Jo in when his birth mother's new husband turned out to be a drunkard. Thirty-eight years from now, when Jo dies and the coroner's surprise at the shape of the body revealed when he

pulls off the jeans and the rough shirt and the men's underwear brings reporters knocking, a trove of letters from the
foster mother will be discovered in a trunk in Jo's cabin, yel-
lowed but kept safe like the precious things they are.

On the train to Idaho, here is what Jo sheds: the years of
growing up, the life before now left to tumble right off the
back of the train like a forgotten parcel. He sheds a family, an
identity, the eight inches of lank dead matter that was his hair.
The *s* at the start of the pronoun *she*. Here is what he gains:
an "e" at the end of his name. That, and . . . everything. He
practices standing differently. He practices chewing tobacco,
letting the sour slosh around his mouth before aiming it out
the train windows, as America rushes by.

In 1867, at now seventeen years old, Joe Monahan steps
off the train and into a new life.

The horse I rode at the stable back then was named Carefree—
the irony lost on me at the time, but too clear now. Carefree:
something we both definitely weren't. He was tall and ornery
and famed at the stable for biting anyone who got too near.
All my orneriness, all my anger and grief, was kept under
my skin then, silent as everything else about me. But I was
tall. I am tall now for someone assigned female at birth, and I
have been this height since grade six, so I was even taller by
comparison then. Maybe that's why the instructor, a lean and
cheerful and to me therefore inscrutable woman, paired me
with this horse, this horse that could only be wrangled and
never appeased. I remember the chestnut sheen of his coat,
the gnarled knobs of flesh on his joints like knots in a tree,

the musk smell of his sweat and the flare of his nostrils. The horsiness of him. I must have felt relief to encounter anything so resolutely itself.

From the start, he had no patience for me. Perhaps he sensed that I didn't have it for myself. He was bored; he was cranky; he had other things to do rather than contend with this stranger with her tentative, uneasy way. Once a week, my mother drove me and my sisters the thirty minutes from our town to the stable, which sat in a park we otherwise never visited. The park was green and fiercely manicured, and from a distance the stable rose as an unruly silhouette, dark and musty and somehow cool year-round and shadowed all the day, the sun's brightness subsumed by the loamy smell of dirt.

First you passed the small tack room, with its rows of bridles and leads, and the spare crops and helmets for borrowing. Then into the corridor where the horses stood in their stalls. I had been hoping to be matched with a horse who would come to know my scent and my footsteps as I walked the corridor, who would nuzzle into me like for once I had been chosen. Everything else was so complicated: the ungainly hoist up to the saddle, a movement that made me contend with my body. The spread of my thighs in those pants, puberty hormones making themselves known. My body became something to be contained. I hated trotting, the neat post-and-sit. I hated, I suppose, orderliness, all the *shoulds* that had suddenly arrived into my life.

I needed something to love. And though he never nuzzled me the way I'd hoped, I found what I longed for every time

Carefree broke free of the trot, shrugged off my cautious reins, and galloped. I clung then and, on his back, flew.

Carefree, I loved.

The year is 1816, and a tall girl named Sarah Dowling who refuses—scandalously—to wear a dress comes knocking at the door of Patience White. "Her breeches didn't hide how soft she is below," Patience will later recall, of this first meeting that begins the classic 1960s lesbian novel *Patience & Sarah*, the meeting that will foretell her fate and bind her to a life as yet unimaginable. "Maybe they even brought it out."

Patience finds she cannot stop thinking of Sarah. And so when Sarah tells Patience of her plan to run off to another county, where she might get out from under the thumb of her family and rent some land of her own, Patience's friendship snarls in the knot of her secret feelings, feelings she doesn't yet have language for. To stop Sarah from going, she seizes the weapon available to her: gender. Sarah can't go, she says. Who would rent land to a woman alone?

"They don't have to know I'm a woman," Sarah says, her hair tucked under her cap, her loose shirt making a straight line of her body.

Fine, Patience says. Then she'll come with her.

On the day that becomes the beginning of the future that follows, Sarah shows up at Patience's house in a dress, having needed to escape any notice from her family. "Don't say nothing," she hisses at Patience, uneasy in the dress. Patience wants Sarah to pass through life unnoticed as different. As the novel's author, Alma Routsong (pen name Isabel Miller), put it in a later interview, Patience tries to teach Sarah "how to act like a lady, how not to hear anything, not to see anything, not to respond to anything, not to put her hands in her pockets, not to cross her legs, not to stride these million little (or huge, really) socially required inhibitions." But even Patience cannot deny to herself that Sarah looks in the dress like "a dog we'd clipped once to help him against the summer heat, and he hid until his coat grew back." The dress is wrong. They don't have a word yet for what is right, but Sarah in this dress is simply wrong.

They have no real plans, they have no money, but they have two cows from Patience. And Sarah's horse. The horse is what they pin their dreams on, what will carry them away from the lives that have no place for them. Poring over a map, they decide on a destination. There, Sarah says, "I'll cut my hair and be a laborer." They never do make it to the chosen town, settling instead in a closer county, but they hardly care. The point is not the destination; the point is the new life: Sarah in breeches, and Sarah and Patience together in their bed at night.

My own dreams were never of the weekly riding lessons sort. Instead they were of freedom. Someday I'd grow up, and I'd move to Montana or Wyoming, one of those big, wide-open states, indistinguishable in my mind but always imagined as a swathe of green with mountains on the horizon. I'd live there alone—in my dreams I was always alone, as though I couldn't even imagine my way out of my loneliness, no Sarah Dowling to show up at my door with her gunnysack and her trousers and her shy, persistent way—but for my horse. It was the horse's chestnut sheen I would picture when I dreamed. I had fuzzy ideas of how we'd live, how it wouldn't run away, where it would go in that valley, and what exactly I would do to earn money out there in an empty field, to keep us in oats and coffee. All my dreams were outward looking. I never quite imagined myself. I didn't even dream my body, for I could not picture myself getting older, because I knew I would have to get older in the body I had. That was unthinkable, so I didn't think it. I could not countenance the distance between what I looked like and how I felt. What I imagined, only, was the wide expanse of grass. And the horse.

Maybe you're thinking of the movie *Desert Hearts* right now, imagining the solitude of young sculptor Cay Rivvers, as she—in a dashing, yet feminine, fitted jean shirt—notes from the back of her horse the arrival of professor Vivian Bell, come to Reno to wait to be eligible to divorce her husband. The two seduce each other, out there amid the desert plains, horses their only witnesses. The movie is a romantic fantasy, a lesbian classic, but it wasn't mine. Those women knew they were women, they were drawn to each other because they were women, and so for me even that story, when I saw it young, wasn't mine.

So picture instead, from Annie Proulx's short story "Brokeback Mountain," and later the movie, Jack and Ennis, friends who can't admit what they are to each other, men who look more like my friends look now, who look more like what testosterone nudges me subtly closer to every day, ever further from that girl in the stable with the body that feels wrong and the jodhpurs that feel wrong. Picture Joe Monahan. Or picture Harry Allen, eventually a legendary horse thief of the Wild West, but in 1908 smiling rakishly as he prepares to ride out of town, to shed another home like he shed his girlhood and in the next town break a girl's heart for the pleasure of it and another and another, and only then will he stop, when, as the story goes, two of his spurned lovers have committed suicide and a third attempted it, their broken hearts evidence for him that he is worthy of love, is alive, is now in his body and can be seen. Picture how much hurt those who cannot see themselves, cannot yet love themselves, can cause.

And picture what a new beginning, a beginning true to one's own skin, can require. What it takes for Sarah to trust Patience with the truth of who she is. Or what it takes for Patience to decide to trust Sarah, to believe that another life is possible, a life away from every *should* and *must* they know. Haven't new beginnings always belonged to the queers? Hasn't the dream of reinvention, of breaking loose from our cares and becoming ourselves, always been ours? And haven't we always had to learn how to do it?

No one knows exactly who was the first to say, "Go West, young man"; the phrase has been the subject of many speculative articles trying to chase down the words that spawned such mythology. But the quote is usually attributed to the newspaperman Horace Greeley, who made his exhortation in the *New York Tribune*: "Go West, young man, and grow up with the country." Greeley, historian Peter Boag notes, was describing meeting a young man who had gone out to Colorado to make his fortune, but had failed, and was headed back home in disgrace. He had later realized, he wrote, that the young man was not a young man at all, but (in the language of the time) a young lady in men's clothing. So Greeley turned the piece into a sexist warning: go West, young man, but not you, young ladies, for surely that was what she was, surely that was why she'd failed.

If this is true, he missed the story right in front of, and all around, him. The one that said you could go West and become who you were, as Joe and Harry and countless other people not at home in the gender assigned to them at birth (many of them documented in Boag's legend-rewriting *Re-Dressing America's Frontier Past*) made themselves a new home on the back of a horse.

When I first came out as a lesbian—back when I thought I would keep the secret that I wasn't quite a lesbian because I wasn't quite a girl—I was often told that I should read *The Well of Loneliness* by Radclyffe Hall as a daringly early example of a lesbian novel, one written in 1928. A classic. I didn't back then; the title seemed so dire, I couldn't bring myself to. What did I need of more loneliness?

So imagine my surprise a few months ago when I finally read the book, and discovered not a lesbian classic, but—we would now say—a trans one. One scene sent me reeling back in memory. Stephen is a child, around the age I was back in those stables. Stephen stands with the neighbor girl, both of them wearing jodhpurs, and the father of the little girl compares his daughter to the other. What that father sees is two little girls. "Ridiculous of course," Stephen thinks to

themself—ridiculous, of course, to be thought a girl—"And
yet all of a sudden you felt less impressive in your fine riding breeches." The breeches weren't the mark of the gallant
young man Stephen had imagined themself to be, the story-
book character called "Nelson," who is daring at the hunt and
daring in the woods. The breeches are, instead—in the father's
eyes, in Violet's eyes, in society's eyes—jodhpurs. And the Stephen wearing them, a girl.

It is not the humans, not the horse girl Violet or Violet's father, who save Stephen. But, rather, a horse: Raftery, a spirited
gelding Stephen has grown up alongside. Stephen's great—
perhaps only—love, on the back of whom a reimagined life is
possible.

If I'd known back then, is what I think every time I discover such a story. *If I'd known. If only I'd known.*

So potent was the cool of the girls at the stable, so much did
the stable seem to belong to them, that at some point I decided that because *girl* wasn't a word for me, *horse* couldn't
be, either. I ceded the space of the stable over to the space of
yet another myth of what it means to be born into a body in
America. We are so good at us-versus-them divides, no place
for the middle. I go horseback riding now only once every
few years, the last time hardly a ride, a bunch of poky old
ponies pockmarking their way across the sands of a Greek
island beach. The view was beautiful: the lapping of the azure
water at the ponies' feet, the bluffs of other islands rising em-
erald and dusky in the distance. The ride was boring. Its only
thrill came when a flock of birds descended on the beach and

the ponies startled and reared up, forgetting their bridles. We laughed and, for a moment, forgot ours. Life has built up around me: I live in a small city now, near the college where I teach, and there is never enough time or enough money for all the things I want to do. Now the cowboy boots I wore for years sit in a big box in the apartment closet, and the only concession to a horsey past anyone would notice is the tie rack in my bedroom, carved in dark wood to make a cowboy's boot spur and a horse's head, approximately the color of Carefree.

But I think back to him often, to the pleasure of forgetting my body on his back when he ran. A historian colleague recently wrote to me, "We historians have a funny way of going about things. We think we can imagine the past, even better than the people who lived it knew themselves." The list of things I don't know about the past is long, even the past I lived. What happened to Carefree? When our time together ended—the end that brought, too, the end of my time in the stable—did his owners just move him away, as we were told in front of our parents? Or had he been sent off to the dog food factory, as we whispered about behind those stable doors? Did he fit in as poorly as I did, as he seemed to—and did our moments of riding give him the ease they gave me? When Joe's horse took off at a gallop, back when Joe was still Jo, and he felt the air rush silken past them, and trusted his body, his muscles, to hold tight, and he trusted the body of the great beast that carried him, did that trust feel like weightlessness, like flight? Did it feel like possibility?

And who was that girl who rode Carefree, who felt so ill at ease in her body and needed him to make her free? If I could reach back across time and give her every story she never knew, every possible future she never dreamed, would I be able to write for her a happier beginning? Or did I need that loneliness she felt, the loneliness that would become a drive to imagine, a drive to bust free?

And will we ever learn what simplified myths cover up? That we are not the only outcast at the barn, that we are not alone in history or in our sorrows. That we stand at the doorway of our own hearts, hat in hand like Sarah, asking our own quiet selves to take a risk and imagine a future.

The girl I was is just as much an artifact of history as anyone I can uncover in a book. Just as mysterious to me. Just as gone. In her place, I imagine Carefree. We do with animals what we do with the past: project our hopes onto them, our human feelings and failings. I used to wish that whoever named Carefree had chosen a more fitting name, a name that wasn't always and incessantly what we both weren't. But what do I know of his ornery heart? I looked, on the outside, like a girl back then, and no one but me knew I wasn't.

He carried some secret of his own, I am sure. I hope he got to live it.

PLAYING SAFE
COURTNEY MAUM

On the fifth day, the horses were released into the wild: three strings of twelve into the scrub. The horses had been in training for the Agua Alta polo match in April and were in peak condition; manes hogged, hair clipped, flanks muscled and twitching. Their shaved manes put them at a disadvantage for their premature vacation. The hot season was approaching, and the flies with it. From our house on the hill, I watched the polo ponies circumvent the horses that had been in the brush some time. These horses pranced and flung their useful manes around while the others hung their heads low, cattle egrets standing sentry next to them. Although I knew how to ride a polo pony and the horses were a short walk from the house that we were sheltering in, I would not be riding them. Nobody would.

I had come to Mexico in early March with my family to promote my third novel, *Costalegre*, which had just come out in Spanish. The trip to western Mexico was the last leg of our tour, a reunion with the seaside village where the book is based. The journey should have been celebratory—I was with my husband and my daughter and had been working hard all year—but while we were in Mexico City, a virus that had seemed contained, or at the very least, remote, when we left, bloomed into a pandemic that was touching the whole world. Faced with the choice between a four-airport journey home plus possible quarantine or isolating in Careyes, a private estate on Jalisco's jagged coast, a long stay in the jungle felt like the safest bet. We called ahead to make sure that Adeline—my father-in-law's ex-wife—would let us overstay our welcome by a week or two if necessary, not yet understanding that we would stay for months. With her assurance, we left the hand-cranked organs and the traffic and the vendors of a city for the bird noise of a savanna and the *panteras* at night.

One airplane ride and a hundred miles later, we reached the wooden gate to Careyes. A guard leaned into our rental car and aimed a thermal imaging laser at our foreheads, calling out the results in Celsius. 36°, 34°, 34°, respectively— these numbers clearly erroneous, because if they weren't, two of us were hovering around 93° Fahrenheit and nearing hypothermia, but they satisfied the guard. We were given a scrap of paper to prove that we'd been gifted entry. Were told not to come out.

Founded in 1968 by the Italian real estate magnate Gian Franco Brignone, Careyes is a twelve-mile-long stretch of sandy coast and tangled selva sitting off a two-lane highway on the Pacific coast. Discreetly housing some of the most fantastical real estate (and fortunes) in the world, this rococo land—where the residents own one-word commodities like "sugar" or "railways"—is something of a muse to me, not only because I'm fascinated by the mores of the wealthy, but also because I'm drawn to the emotional complications of monogamy and divorce, topics our hostess Adeline espoused in

spades. Adeline had turned away from a career as a professional cellist to marry my husband's father, and when his
womanizing sabotaged their union, she took up with another
Italian, the aforementioned Gian Franco, who enthroned her
in a cobalt blue ocean *casita* for a decade until he replaced her
with a younger model and gifted Adeline with a tract of land
as a consolation prize. Referred to among locals as the "ex-
girlfriend plots," Gian Franco's parting gifts to his ex-lovers
explains why Careyes has a curious mixture of billionaire ty-
coons and scrappy, bohemian feminists living in its arid hills.

In addition to it being a source of inspiration for my fiction, Careyes reignited a horse obsession that lay dormant many
years. I can track the deepening of my relationship to riding
with each of our trips here. On our first visit (me, my French
husband, and his Italian father, en route for a two-week stay
with his ex-wife), the car ride from Puerto Vallarta that takes
three and a half hours today took six then. The road was dirt
from kilometer 175 to Adeline's home at 50: eroded and crum-
bling in many places, washed out in others, blocked by cattle
constantly. On that trip, it had been twenty-seven years since
I'd last been in a saddle. The horses we passed in the corrals and
cattle ranches of Jalisco thrilled the horse-mad kid within me,
but they didn't tweak my muscles; I didn't need their backs.

On our second visit to Careyes, it was only me and my
husband, Diego—his aging Italian father having declared
that, *primo*, the journey over was too long for him, and *sec-
ondo*, our last visit to his ex-wife's had reminded him of the
reason that they got divorced in the first place. As for his
son—my husband—Diego was just as taken as I with the per-

verse beauty of the region, and I think he felt an allegiance to
Adeline—a woman who had never had children, partly because her ex-husband hadn't wanted them. It made Diego feel
good—or at least, *assuaged*—to check on her biannually, to
make sure his would-be stepmother was sound and safe.

On that second visit to Careyes, the road was just as terrible and the horses just as pretty, but expensive lawn art they
remained. By our fourth visit, however, I was a mother for the
first time and a rider once again, and I couldn't pass a horse
without wanting to run up to the pasture gate to feel the hot
exhale of its muzzle, imagine what the bush would look like
from the slope of its sweet back.

By that visit in 2018, the road to kilometer 50 had much
improved, and I joined the Mexicans and the Argentineans
who gathered in the jungle to chase after a polo ball on a field
bordered on the southern side by a crocodile lagoon. This
month—an endless month in a century of a year—marks our
fifth return. The people who used to yelp and shout and swear
on the horses beside me have put hurricane protection on
their windows and turned their horses out to pasture where
the boarding was cheaper and the feed was free, signing
themselves—and their horses—with the cross first. Though
I miss the players' companionship, the missing is nostalgic.
I wouldn't be out there with them even if the virus hadn't
come. It took me a long time to build up the confidence to play
polo competently. But something happened to me this year,
and the confidence that used to allow me to charge into a train
of horses with my mallet at the ready isn't with me anymore.

———————

Except for my ongoing obsession with the myth and mythology of horses, there is nothing about my pocketbook or
personality that cut me out for polo. Sometimes brave emo-
tionally, I'm the opposite physically: I sit down on my behind
when I have to descend a steep slope, especially one with
rocks; I'm tired out by tag; I have a twenty-minute stamina
for most physical exertions, including sex. I'm prone to panic
and depression and anxious, racing thoughts. My high-strung
disposition makes the risky sport of riding even chancier for
me. When I mount an equine, they take in what my nerves are
communicating, and they think they need to run.

Although my life today varies between scrabbling together
paychecks from my freelance writing and enjoying the rare
windfalls of a book well sold, growing up, I knew a life of favor
and stability—at least of the financial kind. The lucky first-
born of a Wall Street father, I wanted for nothing, needed for
nothing; my little-girl horse madness was heartily indulged. I
had a brown pony and riding lessons, special ribbons for the
braids I needed to score turnout points at horse shows, leather
garters to buckle round my six-year-old calves.

When I was nine, my parents divorced, and the horseback riding stopped. So did the stability. My father remarried quickly and made new children with a new wife. My
mother married a man who already had some children of his
own. I wasn't an only child any longer, and I definitely wasn't
the favorite; there were new babies and kitchen redesigns to
oversee; big life plans that didn't necessarily include me. It
was around this age that I deemed self-sufficiency a survival
tool, a decision that led me to go after my first of many jobs—

shopgirl in a costume shop—so that I could make my own allowance instead of blushingly asking my father if I could have pizza cash.

The pride that came from a small degree of financial independence (earned six dollars an hour, by hour, for so long) was supplemented by a new fervor for writing. The cracks that had previously been caulked by horses and the security of married parents were mended by invented worlds. Sometimes I wrote about the horses that I considered lost to me, sometimes I wrote about the boys who kept me from mourning the lost horses, but eventually those elements fell away, and it was just the stories. I was building safe, expansive word-worlds that I could live inside.

I tried to fight my return to horseback riding at thirty-seven years of age. The novel I was working on at the time had horse people in it, a dressage rider and a breeder, and I'd started conducting interviews with riders to put flesh on my characters. I assumed I'd be able to fight off the intoxicating barn smell and the high that comes from the potent musk of horses with the armor of my empty bank account. By that point, I'd been a proudly independent but not-so-proudly broke freelance writer for decades, and the price of something like sesame oil was enough to make me wince hours after I'd purchased it. The truth was in my tax return; I couldn't afford to ride.

But at thirty-seven, my center wasn't holding; the cracks inside of me were inching toward a gape. After the publication of my first novel, I became the financial hope for my husband and child, which meant not only that I had to write well and often, but that I had to write commercially. My daughter turned two, and in addition to the caretaker she'd successfully found in me, she suddenly wanted a playmate. I was stressed out, foundering, insomnia-plagued, and overworked; I couldn't still myself to play. My husband convinced me to sign up for a therapy appointment, and after filling out the mass of intake questions, the therapist suggested that I was too depressed to drive home alone.

I tried medication; it worsened my bad sleep. I tried aroma-therapy, supplements, energy therapy, yoga, meditation, acupuncture, journaling. I tried running many miles. My husband and I signed up for couples therapy. I increased my iron intake. I tried eating less bread. I stayed off social networks, avoided the documentation of other people's functioning. On my agent's advice, I took time away from a manuscript that had become a parasite. I missed my book deadline. I didn't get the payment I would have if I'd made it. Lost in a hell of my own making, I barely slept at all. As a last-ditch effort at self-care, I signed up for a riding lesson at one of the barns where I'd interviewed a trainer for a short story I hadn't published. I put a cracking leather saddle on top of a big, black horse who smelled of the hay and neatsfoot oil of my easy childhood. In the ill-lit ring, I pulled myself into the saddle and looked straight between the patient horse's ears. A lost part of me returned.

My six-year-old daughter and I have taken to tracking our time in Mexico by the horses' movements in the valley. At our last count, there are thirty horses grazing—most of them bays, except for a white horse with a black mane and tail that we

have dubbed "The Unicorn." This horse is majestic and powerful and looks up from his faraway perch when we call out.
Our day begins with The Unicorn's approach. Although he
isn't a polo pony—his mane was never hogged—he appears
to be the herd leader, guiding the horses out of the valley that
they sleep in (which we can't see) to lowland that we can.
Around eight in the morning, The Unicorn ushers in a slow
parade that ends with the various herds taking up positions in
the pasture: most of the bays with bays, the white horses with
the pintos, all of them accompanied by their loyal, pecking
birds. At noon, the horses disperse to the shade of the cape
figs to stare glumly into space, motionless enough to scare me.

Until our equestrian espionage in Mexico, I had never seen
so many horses fall still at the same time. *Something worse is
coming*, is what I told myself the first time that I saw them,
dead still in the valley. The next day, when they napped in
the same pattern, I felt ashamed of my suspicion. Around six
p.m., the horses call out to one another and make their slow
way home. There are at least two foals so far, maybe more;
it's possible there are others. So much can happen here in the
jungle while the grass waits for the rain.

Sometimes I think that a propensity for polo was subliminally
drummed into me from the way that I grew up. My father
has been a lifelong wearer of Polo by Ralph Lauren, and I
have strong memories of my sink-high self sneaking into his
dark bathroom to examine the way the liquid tilted back and
forth inside the green glass of the cologne bottle, the smell a

stand-in for a father absent on business trips that were not strictly business.

There were also polo mallets around the house, used as decorations even though my father didn't ride. When my parents divorced, my father moved to a development in Greenwich, Connecticut, that had a polo club inside it. This club was more or less defunct by then, the field more of a depot for the owner's sculptures than an active playing field, but still, on weekends, I would take my bicycle out to that green swath, scrambling up into the empty bandstand to scour the property's tree line for errant polo balls. I never imagined myself riding a polo pony—absolutely not. I was dreamy, prone to spaciness, uncoordinated in a body growing fast. In school, gym class was the subject I feared most. While I was more than happy to toil away over my textbooks, I did not want to sweat—especially at an age where we weren't permitted showers; we were just supposed to exchange our funky gym clothes for the uniforms we'd left heaped on the locker room bench. No, I never imagined myself *that* kind of a rider. If anything, I spent many of those polo club explorations pretending to be the horse.

The Greenwich Polo Club has come back to life today, a destination for the shift-dress and boating-hat set who pay to sit as close to the red sideboards as possible, even though this means they are in harm's way of the flying turf sods and the galloping mares that make them fly. I haven't watched a game there since the club opened back up; my father left Connecticut in 2007, my mother in 2005. The Greenwich Polo Club—

all the polo clubs—will surely lie dormant and bird-filled this summer; the USPA has forbidden grass and arena matches, and even here in Mexico where the governing body is a different entity, the horses are set free.

Normally, around this time in Careyes—the very eve of April—the grooms would be exercising their *patrón*'s numerous polo ponies on the gallop track twice a day, sending up *ranchera* music and dust blooms in an effort to prepare the horses for the Agua Alta match.

I worked these warm-up laps with the grooms on my last visit. You ride one saddled horse and you "pony" three or four other horses, leading them alongside. Two sets around the track at a walk, a trot, and a gallop each, then time back at the barn for watering down the horse's legs where the tendons warm and throb.

Those warm-up laps were some of the happiest moments of my life. To feel part of a moving energy, to be in a line of so many beating hearts, to hit a turn in the right way and keep four horses at heel, the burn of the nylon rope that brought three thousand pounds of horse flesh into the control of my small hand, these were times when I felt powerful and beautiful and capable of so much. It has been true for me in the four years since I've started riding again that I feel happiest when I can control a horse. When I can't, I come apart.

We build responses to the horses that put delight into the day. When we hear the thunder of a gallop, I yell, "We've got a runner!" and my daughter and I dash to the veranda to see which horse it is. There is always a troublemaker—most days it's a dun gelding with a striking black mane. A persistently slow learner, he's always sniffing his way into a herd of mares, running up small hills to cut them off at a narrow pass where he's inevitably bit by competitors and exasperated mares. There isn't a television here at Adeline's, and the internet isn't strong enough to stream programs or watch movies. The herd is our entertainment and our grandfather clock. It is also a form of therapy for my heart and ears; with every neigh that rises up the hill to us, I feel comforted in my identity as someone who writes about horses more than she rides them. I am a child in the bleachers of the polo club again; observing from a distance that feels powerful, but safe.

Enough time passes that the remaining grooms move the horses to a pasture we can't see. It's a question of new grass; even terrain this expansive can't sustain a herd indefinitely. Although the move is necessary, I'm not handling it well. Now that the horses are far away and out of view, the energy that the simple act of watching them provided is leaching from my body. I am limp and lost. Not someone who watches horses, not someone who rides them.

In the absence of the herd, we have started counting arachnids. At least one scorpion goes into a jam jar of alcohol with a pair of toast tongs each night. Adeline has a friend here who bakes them into resin trivets, and he has asked us not to squash them, to save him the big ones that we find.

Normally, we find the *alacranes* hiding in between the wooden shades that cover the mosquito screens in our bedroom windows, but the other day, Adeline woke up to realize she had shared a bed with one. Aside from loosened bedsheets, they like to sleep on top of doorjambs where it is dark and still. We found one of these door sleepers the other day, a royal flush: a huge scorpion made even larger by her pregnancy. Although I know that a scorpion bite would be a disaster in these times—the city hospitals overburdened because of the pandemic, and in the case of some small villages, closed to keep out the sick—it has deepened the growing sadness within me: the pitch of Adeline calling for my husband, the *thwack* of a slapped shoe. They're always sleeping, these scorpions, just sleeping somewhere cool, doing their best to stay safe and sheltered in the dark.

The valley below our house is empty, dry, and yellow. The rain won't come for months. The grooms have to keep on moving the horses around to the places where there's grass. In the absence of running hooves and our efforts to name them, I think about the way that people have yearned for horses and believed in their powers over time. The Romans, the Celts, the Persians—so many cultures have used them as an augury, in battle especially. Entire wars would be called off depending on the way that a white horse walked (or didn't walk) over a spear. It was believed that an oath taken on horseback could never be broken—never ever—no matter what. Another expectation passed down by old wives: if a horse neighed outside of your front door, a member of your family would fall ill. I believe in augury. With the horses' relocation to another pasture, I can't track the hours anymore, nor the terrible things happening within those hours as they pass.

ITEMS DROPPED IN THE JAM JAR OF ALCOHOL SINCE OUR ARRIVAL ON MARCH 12TH:

Six scorpions (baby)

One pregnant adult scorpion

Six medium-size scorpions

One giant desert centipede—a furious biter, found writhing by my husband's foot as we spoke to our French friends about the hearses double-parked on the streets of Paris

<u>Breakfast:</u> Discuss the quality of the sleep we did or didn't have, strategize on what to do with the chachalaca birds that keep us up with their tin-can cackles, consider what we will make for lunch even though it's always the same.

<u>Lunch:</u> Silence.

<u>Dinner:</u> Beer, pasta for our daughter, we don't fight her on trying new vegetables anymore.

<u>Evening:</u> Listen for the horses. Hear cowbells clang, instead.

After my comeback riding lesson at the first barn, I landed at a smaller place in western Connecticut with reasonable group lessons. I was put into an afternoon class with female retirees, kind women who brought post-lesson sandwiches of egg salad on wheat bread. I trotted around in circles for a few months before I admitted that as much as I relished the newfound contact with the horses, I needed a challenge. I needed to shake my soul up and run down the fears that had been clutching me for months.

Maybe my father's cologne imprinted a fetishistic relationship with polo mallets into my head as a young child.
Maybe it's a hang-up with the trappings of the elitist lifestyle
I rejected when I decided to make my own way as a writer,
whatever the cause (mine now to find in therapy), I asked my
instructor if she knew someone who could teach a total beginner polo, and there I was, off to see one man and then another
and another about getting on his horse.

I was youngish and not unattractive; men were generous
in teaching me what they knew about the sport for free or
at a steeply reduced price. Other times I traded copywriting,
press releases, work on a website. At the places that allowed
it, I worked off my lessons physically, exercising the horses,
cleaning stalls and wrangling hay, my forearms furious with
rashes from the straw. I loved the necessity and the monot-
ony of these daily chores, loved scrubbing out the algae from
a horse's water bucket, separating manure from fresh wood
shavings, washclothing the morning gunk out of a horse's
eyes. I found it all methodical and calming, but I felt this way
about it because I was working in exchange for something I
wanted. It wasn't my day job.

Absurd as an athletic endeavor (so dangerous! so classist!)
polo nevertheless came to me—or I to it—at a time when I
needed something that would make me trudge through the
darkest marshlands of myself to an arid other side. Two years
after becoming a mother to a child who initially brought me
pride and a purpose to all hours, I had become a mother who
didn't know how to entertain her daughter, wife to someone

I didn't want to touch or to be touched by, author of a book I couldn't write, dame to a depression whose existence I denied.

Riding brought me happiness and forward motion, but in order to play polo, I had to have a heart-to-heart with the demons underneath my rug. I was an impatient person and a nervous one, I did things quickly and poorly just to get them done. I was a taker of shortcuts and dishonest about my feelings, and while I could get away with these lousy traits in my forgiving group lessons, in polo, they got me thrown off the horse. When I rode quickly after the ball without having put the time in to improve my seat: tossed. When I proved unable to leave my dark thoughts out of the arena, I was run away with, lost my stirrups, struggled to stay on, and all of this in front of teammates who had managed to clear their minds before they mounted. Additionally, I was terrified of speed. Whenever I felt the rush of energy build beneath me I would tug back on the reins. I couldn't get past the conviction that a fast horse equaled a horse out of control. And so I was constantly behind the game, outside the play, losing a battle with a horse in my very efforts to control it. It soon became clear that if I didn't ditch my worst-case scenario attitude, I was going to get hurt.

It took nearly a year of getting on and on again, giving myself pep talks that I failed to deliver on in the arena, months of bad-faith reasoning before something finally clicked. One morning, after warring with a pint-size hotshot named Macarena, I just let her run, and something marvelous happened. I realized that the mare's desire for speed didn't come from panic, but from the desire to *play*. She was a player, and an excellent one at that, and her previous fights with me had been about me holding her back from doing her best job.

After that epiphany, I learned that when I trusted the horses (and thus, myself) to go quickly, the play was tremendous fun, a level of fun I had never before allowed my body. When I let up on the control obsession that was always primed inside me, I realized something else: I was good at polo. I had a natural swing and a clean pass, and although I wasn't great at the rideoffs—one horse crashing into another to throw the opponent off the line of the ball—I managed to be where I needed to be to get the ball that emerged from such a fight. This new realization was baffling to me, it was a crisis of identity; all my life I had thought of myself as a nonathletic person who had nothing to offer except for excessive languor, when buried deep inside of me there was a team player who wanted to laugh, score, and belong. All of a sudden, I was the proprietor of something unbelievable: athletic potential.

This is where my love for polo and the contents of my pocketbook clashed. I did not have the money to exploit that potential. I was worlds away from good enough for anyone to let me ride horses for free in matches, and I didn't have the money to rent horses for the seven-and-a-half-minute match segments known as chukkers. And while I had the money to buy a horse (there is always someone out there looking to give away a horse for nothing), I didn't have the money to pay for a horse's board. And so it was that I found myself at a kind of stalemate; I'd improved up to a certain point and could not afford to improve any more. Big picture, this was fine for me. I was more than happy in my weekly arena matches (which were much cheaper than grass), and in the summers I was still invited to warm up the *patrón*'s polo ponies with friends who worked as grooms. I kind of liked the idea that I had a secret talent, one that would necessitate exceptional circumstances to show off. But then something unexpected happened. My obsession with control returned, and my sadness with it. When I did have the occasion to ride, I rode terribly again.

Week three. Week four. The horses have been moved to a pasture by the beach where sun-whitened crab skeletons dot the land like ghosts. We can't hear them. We can't see them. We eat tortillas every day, for breakfast and for lunch. Even the tortillas are cracking from the sun.

We find more scorpions. The jar is nearly full. A tarantula clings to the front door and my daughter names it Fluffy. At night, the courtyard garden is full of raccoon-like coati who gnaw on Adeline's limes.

The barn below the house stays still and mostly empty. The ocean starts to smell. "It's the red tide," a local friend writes in a WhatsApp message to me, *"La marea roja."* The air cools and the sea thickens with plankton. I wait for whales to come. We walk on an empty beach filled with stranded blowfish, their bodies puffed in death. We come upon an eel with a mouth full of yellow teeth. This could end soon, this fearfulness, this terror. But it probably won't.

A local friend and polo player who used to give me pointers admits that he's been giving lessons to an oil heiress on his ranch. The girl's oil-baron father thinks it's passable to check-box the precautions; the heiress has a driver who wears a mask and gloves, she tacks the horse herself and rubs down all of the equipment with bleach wipes, terrible for leather. But it's been a few weeks, and now the oil baron is worried about what people will think if they find out that the wealthiest community members aren't actually sheltering in place.

But she is hooked, his daughter. She won't stop learning polo. A compromise is eventually landed on; a seaplane touching down. The oil baron buys two of my friend's horses and leases out his groom and they improvise an arena on their oil-money land. "She was getting quite good, actually," my friend says of his lost student, and this makes me jealous. Last year at this time I was playing in his arena with him, and even though the government has told us not to leave our houses and I'm scared of speed again, it stings that he hasn't asked me to return, that I don't have the money or the confidence to play like she can play.

Just like horseback riding, writing is a stallion that needs to be dominated and fed. Stay too long away from it and it will convince you to stay away some more.

But if you avoid what intimidates you for too long, its size becomes prodigious. You'll not approach again.

The year my confidence left me, I was touring for two books. I was far away from horses, sleeping in strange houses because there wasn't a budget for hotels. Never a good sleeper, I was doped up on Ambien to turn off the electricity inside me. In the mornings, I had to stuff myself with protein-rich foods and chocolate to override the sedative stupefying my brain; a necessary fight back to alertness so I could teach and tour and interview, shilling for my books.

During those months away from horses, I went from missing them to dreading the moment that I would ride again. What had been a well of calm for me soon became a bog; polo was a depository for my worries and self-fulfilling deceptions, the things I didn't like about myself, the feats I wasn't up to. In November of 2019, I was home long enough to join my arena team for a match. It was cold and the horses were under-exercised. They were dancing with nerves and excitement in their boxes, and when my coach told me she was putting me on a horse from her personal string that she called "The Machine," I was too flattered to voice the fear inside my heart. My worst-case scenario attitude returned, I convinced myself that the horse—already fast in normal times—would be unstoppable in the cold, that I would be too weak for her and sloppy, and this is what occurred. The Machine was influenced by my terror and tore around the ring, head high and rear ready. I eventually lost my stirrups, so she went faster, still, further panicked by my change in balance. It ended with the groom and the coach forming a wall of their own horses so that my horse would stop, and with me convinced that I was worthless, had dreamt up all my progress, should not ride again.

We haven't seen The Unicorn in weeks, now. From our perch on the hill, I can see my friend Melanie arriving and leaving every day at the defunct polo barn below us, carrying on the same schedule—feed, water, and ride at 9:30 a.m. and 4:30

p.m.—that she carries out each day, a decade of such rituals stretching out behind her. She's the only rider who has continued with her practice during COVID-19, who tacks her mounts herself. I have a text message from Melanie inviting me to ride; she knows that I have quarantined, that I'm playing safe. When I close my eyes, I can feel the girth of the retired polo pony she lent me to accompany her in past years. His name is Poker, and he has sleepy, gentle eyes and a demeanor she calls "noble." But he's fast on the beach. I haven't responded to her message yet. Instead I ask myself, is he too fast for me? When I walk with my daughter on the beach in the cool hours, I find hoofprints in the sand. Proof of Melanie's movements. Proof of her consistency.

From the house that isn't ours, I look at the azure horizon that slips into the Pacific, a vastness so great it can hold both hopefulness and dread. If I don't get back on a horse, I am going to become depressed again, a prisoner to my fears. If I fight those fears, I know what I will feel like; I will feel free and more alive. Because it isn't dead yet, the small moments of greatness and the impractical potential that I carry low. I am someone who can hit a ball one-handed at a gallop. Twist over a horse and hit the same ball backward at a gallop. Fall off a horse at forty and get up to run again.

I reply to Melanie. I thank her for the invitation. Yes, I say, I'll ride.

THE SHRINKING MOUNTAIN
NUR NASREEN IBRAHIM

As a child, I liked to sketch my version of a piece of art over and over again, copying it from an old printout. Two radiant blue horses locked in a playful prelude to a leap. One horse curves away from the viewer, its back leg lifted, the muscles in its neck straining as it twists. The other faces forward, its mouth open in what looks like a wild exclamation of glee, its front legs floating above the ground.

The horses could be suspended in water, were it not for the stark shadows beneath their feet. From a distance, it looks like an old, crumpled photograph. I must have been around twelve years old when I studied a printout of the horses closely one summer and observed small cracks I'd never noticed before running through it. What I had previously seen as fluid was revealed to be a mosaic of rich blue stones, ranging from the colors of the night sky to white marble.

Those are the horses of my dreams, the horses depicted by the artist Ismail Gulgee. The first time I saw them was in a book about Pakistan's greatest artists, commemorating Gulgee's career with pages that were filled with his signature calligraphies. They stood out in brilliant tones of aquamarine and light blue, amidst the multicolored oil paintings that the artist was most known for.

The image had a rough-hewn look about it, and multi-dimensionality that made it rise above the smooth and shiny pages containing it. I would press my fingers on the pages, perhaps hoping to feel the roughness, to grasp the sense of a homecoming it awoke in me, even though I was home, I had lived in that same home my entire life. I could not explain it, but I understood it.

My mother explained to me that this was not a painting. Gulgee's horses, both wild and in the midst of polo games, were immortalized in lapis lazuli, stones that traveled from mines in Afghanistan to Pakistan. These rocks came in shades of rich blue and turquoise, and were so rare and highly prized that the artist used them for portraits of world leaders.

My attempts at sketching those horses in my cheap notebook always went awry. I would glance down at my work, which looked nothing like the crumpled printout. My horses were ill-proportioned, their legs too long, necks too short, whereas Gulgee's were elegant and proud. But still, in a small way, I felt close to the artist.

My family had known the artist for years, but the first time I recall visiting his studio was in 2003, the same summer I attempted to sketch his horses. Ismail Gulgee's studio in the mountains was minutes away from our summer home in Nathiagali, a resort town in the mountains of Pakistan. During those long, unsupervised days, my cousins, brother, and I traipsed onto the road above our house, crossing lines of horses bearing tourists, the dark pines, locals grilling corn over coal and crackling fires, past the waterworks and the broken-down building that managed electricity for the hill station, and arrived at Gulgee ki Aankh (the Eye of Gulgee), the unofficial name given to the artist's residence in the hills.

He resided here almost every summer into the early 2000s. The portion of his house visible from the road was a puzzle of mirrors, rocks, tiles, and beads that formed a large blue eye nestled under a triangular tin roof—hence the nickname. The eye watched us as we walked past. Some days it shone in the rain, or from the condensation in the clouds sinking around the tin roofs. Sometimes the eye looked tired, the cracks in every piece of glass that made up the iris and cornea appearing sharp in the sunlight. Appointments with the artist were rare. But on the special occasions when we were able to secure one, he welcomed us into his studio.

The first time we visited, we were firmly instructed by whichever parent accompanied us to sit in a line and not touch anything. I folded my hands tightly in my lap, my stomach fluttering with anticipation, and we all sat obediently on rickety stools, sneaking glances around us.

A diminutive man in a paint-stained shirt that hung over his withered body, he greeted us in a rasping voice, and paced in front of the blankness of his canvas. In those days, approaching his twilight years, Gulgee painted large abstract calligraphies, focusing on Quranic verses or the ninety-nine names of Allah. We took in Gulgee's canvases, the glint of tiles strewn in a corner, the splashes of color on every surface, and the brushes larger than our hands. Next to his easel was a chipped china teacup (that we all assumed held alcohol), which he would gulp down before facing the looming canvas.

Gulgee's body appeared to expand as he confronted the

emptiness before him. His brush, already coated with a variety of colors from the messy palette, quivered in his hand. From my spot on my stool, facing the same direction as Gulgee, the space on the canvas looked like a challenge.

And then with a jerk and a leap, catching me by surprise, the old man transformed into a dancer. In the blink of an eye, his small figure in his stained white shirt dashed across the canvas twice his size, twisting this way and that, with shocking flexibility. Before we knew it, Gulgee returned to us, the same soft-spoken, smiling, hunched-over figure; and behind him, a riot of colors told stories in every curve, of faith, histories, myths and legends, and places that existed only in his imagination.

Much has been written about Ismail Gulgee, and I had a rare glimpse of the genius at work. Yet I knew so little of what really was going on. Watching Gulgee paint, I felt envy, admiration, and an inadequacy that continued to plague me throughout my life.

In 2007, Gulgee, his wife, and a domestic worker were found dead of asphyxiation in his home in Karachi. The police suspected their chauffeur and another member of their domestic staff of the murders. Though the evidence pointed directly at them—they were caught with Gulgee's car, a camera, and two of his paintings—the case lingered for almost a decade in court. Eventually, the two men were sentenced to life imprisonment.

I think often of the artist, who once insisted on gifting one of his paintings to my mother, saying, "You're like my daughter, just take it."

What could have happened that made those men kill such a kind man? This is where my story stalls, limited by my imagination and the biases ingrained in me from my childhood.

Of course, questions grew, from Karachi to Nathiagali's well-heeled community of summer residents. I overheard comments that meant little to me as a child, but took on new meaning in adulthood. The murderers were footnotes, two aberrations who cut short the life of a great man.

"They just wanted some money, and something went wrong."

"They were tempted by the paintings."

"A typical crime story where employers are killed by the servants for money."

One of Gulgee's paintings could be sold for millions of rupees, or thousands of dollars, a few years of a good life for someone who has nothing. And as such stories go, Gulgee's killers languished forgotten in a prison cell, while Gulgee and his paintings were immortalized.

Whose stories are we telling? People like Gulgee—people like me—self-appointed storytellers are often held back by the limitations of our gaze. Gulgee's paintings immortalized princes and presidents, tyrants and humanitarians alike. Perhaps his gaze, like mine, chose not to look at the troubling layers within ourselves, the ones we carefully cover up with time.

Living in New York City today, I still frequently look at images of the horses he chose to portray. Some of them are wild, and some are pressed up against each other in the midst of a polo game. Underneath their majesty is a frantic dark energy that sets my skin on fire, because I can't imagine ever being that free, that brilliant, that reckless. The world that surrounded me when I was a child, during those summers in Nathiagali, with its military orderliness, its quietude and beauty, carried the same darkness; only it was invisible to me.

My cousins, brother, and I rarely planned our destination when we set off on horseback every afternoon. There were not many to choose from. Sometimes we took the pagdandi, the wooded, broken pathway down to the Governor's House, a remnant of British rule on the Indian subcontinent. Other times, we took the road that twisted around the mountain and that also ended up at the Governor's House.

Like all city folk, my family visited during the summer, when snow did not block the roads, confining villagers to their homes. Escaping the extreme heat of the plains, we emerged in the cool mountain air, availing ourselves of the local bazaar and hiking amongst the pines. Nathiagali is a small holiday township, surrounded by a smattering of villages and situated on one road that circles the mountain. A few offshoots branch into walking paths or clusters of old summer homes. On sunny days, when looking for a challenge, my family and I hiked the fearsome Miranjani mountain, its tip visible from every vantage point of the hill station, or the smaller Mukshpuri known for its fields of daisies and a mythical witch who was said to have haunted its pathways.

We began to frequent this sleepy hill station in the sixties when my maternal grandfather, an officer in the Pakistan army, brought his children and grandchildren to the numerous cantonments that populated the region. I was introduced to Nathiagali for the first time in 1991, when I was six months old. We drove up from the city of Rawalpindi, where I was born in a military hospital, and where my grandparents still live. In one photograph, I squint angrily against sunlight and bitter cold, my head wrapped in a wool cap. My father, holding up my tiny body, is kissing my cheek, his prickly beard no doubt irritating me.

In 1996, my uncle had a summer home built for the family on the edge of the mountainside. The house, perched along a winding road, used to be filled with children, cousins, aunts, and uncles, often staying four or five to a room for months at a time. Fourteen years later, my parents built another house on the same road. My father's law practice was flourishing, and by then we were firmly entrenched in the group of affluent families with property in the hill station.

Our house, which is still there today, is made of gray stone, with a green tin roof and a long verandah snaking around its exterior. It sits on the edge of the mountain overlooking a forest. Beyond the verandah, a steep drop filled with pine trees that slope down, down, down. Our front door opens onto a small winding road. One end of this road is at the Pakistan Air Force cantonment, and the other direction leads to the heart of the hill station, toward the bazaar, the historic church, the park, and more homes.

I was six months old when my mother first carried me on horseback in Nathiagali. The horse's owner, Shakoor, came from the nearby village of Namli Maira, and for the remainder of our childhood and adult years, we would not ride on anyone else's horses.

These were horses for day-trippers, regulars like us, and hikers who wanted to go to Miranjani or Mukshpuri. Shakoor was the designated leader of a coterie of his brothers and friends. Maqsood, Mustafa, Pervez, all young men with the same sunburnt skin and infectious grins, would guide us up and around the hills on horseback. Rival groups of ghoray walay, horsemen, roamed the area, but my family was loyal to Shakoor's crew, who had been taking them around the mountains for decades. They gave us unlimited riding times, and we talked them up to other visitors to get them more customers.

As I got older, my cousins and I were dispatched every afternoon with Shakoor's group of horsemen to roam the roads and paths. I was usually last in line, the frequent butt of jokes, and always quick with my tears.

Offhand comments about my horselike nose sent me on a spiral of self-loathing. I believed I was undesirable, awkward, and unsociable. I had a nose too large for my face, eyebrows that were too bushy, thin hair that remained cut close to my ears. The girls surrounding me, all my cousins, had long, thick hair, small delicate faces, narrow noses, and wellsprings of confidence. We loved each other excessively, fought with even more fierceness, and focused our emotions on the validation only girls can provide each other. The harsh words we often exchanged left marks on my ego.

Self-loathing also appealed to me in an odd way. To hate myself excused me from upper-class guilt, it excused me from doing anything about it. It made me focus on my studies and refuse to rebel. But in Nathiagali, I could escape a world that fed that self-hate.

Back home in Lahore, I studied at a private all-girls school, where schoolteachers constantly reminded us about the breakable nature of a reputation. News about a girl spotted in a car with a boy would make its way into the school staffroom, where, over cups of tea, teachers would analyze the failings of her parents, her entire family background, and all the decisions that led to the fateful day she got in the car with a boy.

Nathiagali was the only respite from the stifling heat and gossip of Lahore. Boys and girls from big cities like Lahore, Karachi, Islamabad, and Peshawar also collected there in the summer. How easy it was to walk outside and bump into a large group of kids, or invite people to a bonfire in the backyard of my aunt's house. And so, I interacted with more young boys there than I ever did in the city. But even there I was too shy, and attached myself to my cousins who were able to navigate all social situations with an ease and confidence I only began to manage as I approached my thirties.

In our early teens a group of us girls would hitch rides in the back of construction workers' trucks because we were too tired to climb uphill to get home. We bought mini-firecrackers from the market past the air force base, in boxes inexplicably decorated with images of Jack and Rose from *Titanic*. If we arranged them in a circle with their wicks facing each other and lit one, the rest exploded in loud pops that frightened the tourists walking past.

But nowhere did I feel a deeper sense of calm than during

the daily ride spent on the back of a horse. Lulled by the steady rhythm of a horse's trot, the sound of wind brushing through the vast branches of the deodar trees like a stream of water, the heaving, breathing mountains that rushed past me the faster I rode, I momentarily forgot my anxieties, petty arguments, insecurities, and inadequacies. And for a time, the feeling of freedom even helped me let go of guilt, even as that freedom came with a history.

After the East India Company established their foothold in India during the eighteenth century, many British commissioned paintings from Indian artists, portraying their idyllic life in this new world. In one such painting by Shaikh Muhammad Amir, an English child with her face covered by a bonnet is seated on a pony, surrounded by three Indian servants, one of whom holds an umbrella over her head. In another painting by an unknown artist, British officers and their wives take refreshments at a table, waited on by Indian servants.

The British re-created this life in Nathiagali. They arrived from administrative centers like Peshawar or Lahore, across the northwest of India, and enjoyed tea in wooden cottages overlooking manicured lawns with magnificent views of the Himalayas.

After the subcontinent was violently partitioned in 1947 and Pakistan was born, our military inherited the rituals, structure, and rigidity of colonial life, and my mother's family inherited a military life. My grandfather was in the British Indian Army before independence and brought a British sensibility to his children, who studied at schools set up by Christian missionaries. My mother's family spoke English with a British tinge, settled with ease into the emptied cottages that once belonged to their colonizers, and gathered in the afternoon for tea just as the British had done before them.

I realize now that I spent my childhood trapped in personal grievances and anxieties, mired in a selfishness that blinded me to the role my family played in larger events, leaving unacknowledged the privileges that were handed to us as well as those that were hard won.

Amir and similar artists specialized in painting commissions for officers of the East India Company, creating visual representations of their numerous possessions, lifestyles, and the servants who surrounded them. But we were now the creators, the commissioners, and the centerpieces of the idyllic paintings that were once presented for the white man's gaze. Groups of servants still circulated around the little girl, but now we could see her face, and her skin was brown.

The military jumped so firmly into the power vacuum left by the British that if you threw a rock on a piece of land in Pakistan, they likely had some claim to it. They ruled Pakistan openly for a few decades, and more recently as a quietly powerful big brother managing an inept civilian democracy. And they carried on the work of their colonizers across Pakistan, with my grandfather as one tool in their vast machine.

Nathiagali played a small role during a key moment of history. One story goes: in 1971, Henry Kissinger was brought to the hill station to recover from a mild case of the "Delhi belly" during a visit to the subcontinent. This story was a lie, part of Kissinger's cloak-and-dagger efforts to bring the

United States and China to the negotiating table. At the height
of tensions between East and West Pakistan that later divided
the country into Pakistan and Bangladesh, Kissinger arrived
in Rawalpindi to meet with the then president, the notorious
alcoholic General Yahya Khan. Khan reportedly suggested
that Kissinger recover from his fake illness in the nearby
mountains of Nathiagali. "In Rawalpindi, we disappeared for
forty-eight hours for an ostensible rest (I had feigned illness)
in a Pakistani hill station in the foothills of the Himalayas,"
Kissinger wrote in *On China*. He instead caught a Pakistan
International Airlines flight to Beijing, where he met for secret
talks with Chinese Premier Zhou Enlai. Meanwhile, the West
Pakistan–led military continued its assault on East Pakistan in
an attempt to suppress their independence movement, resulting in millions of Bengali refugees flooding into India, mass
rapes, a war with India, and a genocide, which goes unacknowledged by Pakistani rulers to this day.

My grandfather, an engineer and colonel, spent one week
in East Pakistan before war was declared between India and
West Pakistan. He was there to report on the state of the army's equipment and was sent back when his orders changed
and the war began in earnest. I found it difficult to ask him
about a time that so many Pakistanis have effectively erased
from memory. Khan was a drunk, my grandfather told me in
his soft voice, and made many poor decisions.

I felt deep relief when I learned that simply by virtue of
chance he wasn't an active participant in the horrors that were
taking place in East Pakistan. But I also wondered about the
little ripples that extended further than our eyes could per-

ceive: the equipment he oversaw, the guns, vehicles, the re-
ports he handed over to his superiors, all part of the vast and
cruel machinery of war.

What a luxury it can be to question our past, without understanding the role our passivity played in those dark histories. The same military that continued its colonial legacy,
that committed unspeakable crimes against a population and
suppressed civilian democracies, also introduced us to a home
in the mountains and granted us good fortune.

The horses approached our house each afternoon, the clip-
clopping of hooves growing louder and louder, the soft tinkling
of the bells on their bridles transforming into a cacophony of
sound. They were a small group of varying sizes and colors,
some bedraggled and diminutive, others proud and imposing.
They had bells, colorful beads, and ornaments dangling from
their bridles and around their mouths and ears. Pink, yellow,
orange, and blue balls of woolen string hung under the horses'
heads like fat dandelions twining around their long necks.

We raced up the stairs outside our front door and were
confronted with these majestic beasts shaking their manes,
grunting, stomping, like large cars revving up their engines.
Shakoor, with his big smile and high voice that emerged from
under his full mustache, called out to us.

Shakoor was usually one of the first faces one saw upon
driving into Nathiagali. We'd find him perched on the edge of
the road, on one of the stone barriers erected at curves to block
cars from careening down the mountain. His brown kameez
fluttered in the breeze, and the tips of his feet in their dusty,

worn-out sandals or sneakers, depending on the day, curled over the edge of the barrier as he peered out at the mountains in the distance. He was confident, even so dangerously close to the edge. He and his brothers had the ease of those who have traversed these hills and mountains their entire lives.

I assumed that all the years spent with Shakoor and his brothers meant I knew them well. But our communication was limited to the daily hour spent on horseback, where we focused on navigating each path, trying to overtake each other, laughing when someone's horse stopped, lifted its tail, and let loose a stream of dung. Shakoor answered our questions about each horse with the patience of someone used to repeating himself to naive city folk, who was aware of the power dynamic that shaped our interactions long before I was. But we never truly spoke to each other.

As I grew older I understood that my connection with Shakoor boiled down to memories that grew dimmer the longer I spent away from Nathiagali. I called him a few times this year, knowing he was frequently ill. After I identified myself, he launched into an excited stream of questions, asking after me, my parents, my grandparents. Shakoor was bemused when I told him I was writing a story about Nathiagali, horses, and my childhood. I pressed him on details about his own childhood, questions I should have thought to ask years before.

Shakoor's father was also a ghoray wala. He took tourists up and down the hills for an hourly fee, relying on his earnings and any tips he made to raise his many sons. Shakoor was just a little boy when he first took up the reins. His father used to strap him onto the back of the horse with rope to make sure he wouldn't fall off. Still, he fell frequently, but the more he bruised his knees and elbows, the more he loved his horses.

He reminded me how I often ended up with the horse named Kajol. She was a small, gray filly with sad eyes. Or Bubbly, a larger, brown, warmhearted horse who always hesitated before setting off on a trot. Bubbly was one of Shakoor's favorites; he could leave any of us unattended on her. The horses who ran with them had names ranging from Bollywood stars to royal titles, to some that made no sense. Prince was a brown horse who didn't know when to relax. Then there was Shehzadi, and the majestic gray-white Ruby Aeroplane.

We often fought over who would get to ride Ruby, and my eldest cousin usually won. I asked Shakoor how he chose Ruby's unusual name. "Ruby is a lovely name for a female horse," he said. "And she was as swift as an aeroplane!"

Our only guarantee of safety was Shakoor's comforting presence or one of his brothers grasping our horse's bridle as we went on our evening walks. We had no helmets, no gear normally found in riding schools in the cities. We wore running shoes and held sticks fashioned from tree branches, some of which still had a few leaves attached to the end. The horsemen had even less.

The narrow roads and paths were either muddy or too smooth, and the horses' hooves squelched in the puddles left from that summer's monsoons. Anyone with a modicum of caution would imagine that we avoided cantering, let alone galloping. Yet that didn't stop my brother, my cousins, or any of the other older boys or girls we spent our summers with. If they spotted a clear road, without indication that any cars were coming around the corner, they dug their heels into their horse's flanks and set off at a gallop. Those of us left behind heard the shouts and the laughter of the horsemen who chased after them, their legs carrying them almost as quickly as the horses.

I was more closely aligned with the little British girl on horseback from Shaikh Muhammad Amir's painting than I realized. Shakoor's responses to my questions were matter-of-fact, some brimming with affection, some simply humoring me, but all tinged with a wary distance that I recognized now, tied closely to our vastly different backgrounds. With a willful lack of awareness, I asked him, "Did you ever want to do anything else?"

He laughed. "I've never known anything else, or thought about it."

In 2012 or 2013, during my college years, I returned to Nathiagali after two years away. I had spent previous summers at various internships, traveling the world, racing after a vague idea of a career that my Ivy League classmates had long been plotting.

I was strolling with my mother along a winding road leading to the church when Shakoor appeared around the bend on a cantering horse, a beautiful white beast I hadn't seen before. His eyes wide in excitement, he pulled his reins and the horse stopped next to us with a scrape of its hooves against the road.

"Nur bibi! I haven't seen you in years!" he said.

"I haven't been here for a few years," I said. "How are you and your family?"

We exchanged pleasantries for a while, and then he asked, "Well, shall we pick you up for a ride soon?"

I hesitated, demurred. I was busy these days. I was more of a walker. His face didn't lose its smile, but something shifted in his eyes. It was a quiet realization that I was no longer the little girl who loved horses.

The truth was that I dreaded spending an hour searching for things to talk about with the horsemen. I was afraid of walking in silence. But I didn't even want to canter off on my own because the recklessness of all those childhood escapades had been seemingly stamped out by adulthood. I now feared injury, I feared looking foolish, I feared losing control of the horse, and worst of all I feared all this was a consequence of womanhood. This time Nathiagali could not erase all my years in cities, navigating judgment and scrutiny, and ideas of correct behavior. Even if I believed I was above it all in principle, my self-consciousness had embedded itself far too deeply in me.

Often after a long hike or horse ride, my head spun from the thin oxygen on the mountain. It was an effect that would fade the longer I spent in Nathiagali, and after downing some orange or mango juice. Whenever I became short of breath, I paused to look up at Miranjani looming over the vista, a benevolent and almost perfectly shaped pyramid. The snow-capped Himalayas beyond the peak merged with white clouds and the bright blue sky until they were indistinguishable.

As I took the view in, the mountains appeared to shrink away from me, almost folding into themselves. If I took a step forward, the mountains moved backward. Science—or, as my father would argue, a trick of the tired mind. Whatever the explanation, this shrinking away also made everything about this place feel more mysterious, inexplicable, unattainable.

Gulgee ki Aankh was demolished years ago, the chaotic eye made of broken mirrors, glass, beadwork, and colorful stones replaced with an empty plot of land. We can no longer catch a glimpse of the legendary artist in his studio, darting bird-like from one side of the canvas to another, or hear his raspy laugh.

I often wonder about the men who worked for him, the ones who took his life and absconded with his paintings. What desperation drove them to commit such a shocking crime? And I thought about Gulgee himself; what advantages did he have, who did he serve, to achieve such success?

Without its occupant, the Eye of Gulgee was without emotion, art without a soul. Now that Gulgee's home has vanished along with him, it is difficult to imagine he was ever in Nathiagali to begin with.

There are many voices left out. We, the ones who put pen to paper, or brush to canvas, are often the ones circling power. Within the margins of every story I encountered, were hundreds more. There were no contrasts, no complete pictures that could show the worlds around us in the clear light of day, only layers upon layers to be unearthed.

Over the decades, Shakoor and his brothers' faces reddened from long days in the sun. Their wrinkles grew more pronounced. Shakoor now lives in Abbottabad, only two hours away from Nathiagali by car. He is training his sons to follow in his footsteps. They avoid Nathiagali over the winter, as most of the roads are blocked by snow. When the snows melt, his sons take their horses from Namli Maira to Murree, a larger hill station nearby, and to Nathiagali on the weekends. Rising costs followed by the pandemic began cutting into daily wages. Shakoor arrived in Nathiagali late in 2020, almost halfway through the summer, recovering from a long illness.

Ruby Aeroplane made it through our teenage years. The horsemen had to purchase new horses, after others got old or injured, a necessary expense that grew steeper every day. In the early 2000s, a horse like Ruby cost Shakoor 7000 rupees, which is around $44 in today's money. Back then, most horses cost the equivalent of $15 or $18. Today, Shakoor told me, a horse bought from Lahore or Peshawar would cost them more than the equivalent of $1,200.

Once, I asked Shakoor what happened when he realized a horse could no longer survive. "If the horse's leg breaks, we have to sacrifice it," he said simply. "We bury them ourselves. Once, three years ago, another horse's stomach was hurting. It was in so much pain that eventually it died. But we have to keep going, for the sake of our livelihood."

"Seeing you kids in Nathiagali is like seeing my own children after a long time," Shakoor always said when we arrived. "You are our own people."

Shakoor's affection for us appeared to transcend the years, my adulthood, and the slow realization that I had grown distant from this world. I was too conditioned to accept our apartness to connect with someone who had spent hours patiently holding my horse's bridle and walking me up and down the hills.

What I understood to be Shakoor's affection could also have been his instinctive need to keep the people responsible for part of his livelihood happy. Could he have allowed himself to be grumpy, impatient, frustrated, anxious, while he managed a group of rowdy rich children on horses? Could I be the person to ask him that? Could ours ever be a true friendship, when one person was holding the reins, and the other was just following in their wake, awaiting a daily wage?

Annie Dillard describes the transition to adulthood like a rude awakening. Children wake up, she says, and "discover themselves to have been here all along; is this sad? . . . they wake like people brought back from cardiac arrest or from drowning."

I don't remember my awakening. I just know it came in small blasts of understanding over sleepy days that meant nothing. What I once thought was a painting of horses was a painstakingly gathered puzzle of tiny stones, placed so tightly together that even their cracks, fissures, and flaws were only apparent when I looked closely. Awakening involved learning that this beauty, this life, was conditional and inherently unfair. But it was also a call to free myself of ignorance, and to confront the truth of my own story.

I enjoyed horses, not because I felt we understood each other, but for the surroundings and people that came with them. Perhaps that's why horse riding is an activity so firmly imprinted in my past and not my present, so much a part of the sacred days of my childhood summers and not the messy cities I still encounter every day. I don't want to change my memories of it. I don't want to replace Gulgee's turquoise imagination with cold, gray reality.

So many changes, and yet so little has truly changed. My young nieces and nephews continue to ride Shakoor's family's horses. Even though our visits to Nathiagali are for shorter periods, and infrequent, we still greet each other with joy. My generation no longer has months to spare in the summer, but someone in our family is always here. We are scattered across the world, building lives elsewhere, but always manage to return.

But these mountains, and their deeper histories, hang on me heavily. I wonder if I have failed to get to the truth of it all, if I am still half aware, with the tricked gaze of a tired mind.

Shakoor's favorite place to relax is by the small wooden church across from a park in Nathiagali. This church has been painted over numerous times, from its original light brown wood to dark brown to an ugly orange, and now back to dark brown. As kids, we held fake weddings inside, pretending to marry our friends and our crushes, ran along the pews, ate ears of corn on its grounds, and played Frisbee and cricket in the park below. Here Shakoor tied his horses to the fence and settled down on the long barrier by the side of the road where one has an uninterrupted view of Miranjani and the Himalayas. From this perch, the longer one spends staring at the mountain, the farther it recedes.

TURNOUT

C. MORGAN BABST

My mother strides across a pasture, holding a jacket in one hand, a horse in the other. At the end of an arc of long-line, the Arab is in midleap, his left hooves aloft, his right legs planted, all four topped in white socks that gleam from careful brushing. My mother's face is open, asking the horse something, and his eye is on her, answering, though his head tosses, throwing mane.

It is 1964, my mother a young teenager in bell-bottom jodhpurs. She and that horse are bonded so closely that neither is at all afraid of the other, though one of them carries a whip, though the other's hooves flash in the air. The look that passes between girl and horse is respect itself. Each honors the power of the other. Each chooses trust, instead of fear.

I must have gasped when I found the photograph on a recent Sunday visit to my parents' house, hidden in a shoebox under the bookcase, because my mother looked up from where she sat, cross-legged, on the floor.

What is it? she said.

As she reached out for the picture, old Fox Photo envelopes spilled from the lap of her bathrobe. There was my brother, a naked baby dancing on the patio table. My parents in their Mardi Gras costumes, circa 1984. A trip to a beach in Georgia, where I hunted sand dollars in a floppy cotton hat. Gris Gris, the first horse I ever rode.

Look—I handed her the picture. *It's Ruzon.*

Oh, Ru. Her eyes grew misty as she took the photograph in her hands, fifty-five years after it was shot. *We loved playing like that. We had so damn much fun.*

And maybe that's what got me: her joy, the toss of Ruzon's mane like a manifestation of equine laughter. My mother loved that horse for his heat, the flare of his rebellion, loved him so much she still mourns his death, by colic, during the first year of my parents' marriage, forty-six years ago. Though the photograph's colors have faded to sepia and beige, her hair and her horse's coat glow the same deep auburn. They are a perfect match: two wildernesses bound by strict upbringing, unleashing one another.

Somehow, though, I'd never seen this picture before. On the bookcase, professional show ring shots and yearling portraits of all her horses stand in frames. In each photograph, my mother's back is ramrod straight, her cheeks bright with blush and exhilaration, the horse's shoulders streaked with lather. Someone, I know, is crouching beside the photographer as the shutter clicks, cracking a whip, throwing handfuls of green shavings into the air to make the horse prick his ears.

I haven't ridden—really ridden—in quite some time. For the first decade after college, I didn't have the wherewithal; besides, I was living in New York, and trail rides in Central Park will get you only so far. Since moving home to New Orleans four years ago, we've had to put down my sweet hunter, Nate, who lived to the unthinkable age of thirty-six, and Pecos, my rescued Quarter Horse, who dislocated his hip in his pasture. I miss them—that's part of it—but really, I haven't felt the drive. At some point, what I loved about riding morphed into something else, became all about control, rigor, competition: things I have enough of, already, in my life.

Still, my mother and I have begun to give my young daughter riding lessons. Standing in the middle of the round pen at a friend's barn, we shout, *Up, down. Up, down,* while she trots a borrowed pony. That she would learn to ride was never in question. In my family, girls grow up to be horse-women, almost by default. Looking down at the photograph of my mother in all her glory, I suddenly remembered why.

I called down to my daughter, who was doing cartwheels at the bottom of the stairs. I needed her to see the picture, to understand the freedom my mother shared with Ruzon when she was a child. As the three of us sat together on the carpet, I asked my mother to tell us the stories I already knew by heart, stories of a girl and her horse. How she and Ruzon ran flat out along the top of the levee, faster than the cargo ships on the Mississippi. How they trotted down St. Charles Avenue to her convent school so that a priest could bless him on St. Francis's Feast.

The nuns wouldn't let him in the chapel, my mother said, *so Ru stayed in the courtyard and ate up the azaleas.*

My daughter rang with laughter at this picture-book perfection: A horse with a mouthful of flowers! A horse who tried to go to church!

This, I thought, is where I want my daughter to begin.

———————

Up, down. Up, down. Up, down.

A trotting horse moves her legs in diagonal tandem. Left fore moves forward with right hind; right fore moves backward with left hind. This motion is comfortable for the horse—she can trot, with great efficiency, for much longer than she can canter. But each doubled footfall jolts the rider from the saddle. If she does nothing, she'll be bounced into the air, her teeth will clack. She might even wind up on the ground. To sit the trot takes strength and continuous, concerted work. Otherwise, the rider must post, using her own legs to rise and fall in rhythm with the horse.

The horses my mother, and then I, grew up riding, though, were built for comfort. We rode Saddlebreds—animals with high tails and even higher knees who had been bred and trained to be both stylish and smooth. These were planters' horses, creatures designed by and for men who wanted to master everything in sight—land, animals, women, children, other men—so that they could ride, in long coats and hats, for long distances, across their plantations and along the dirt roads into town, and then parade down city avenues in style. To the Saddlebred's natural walk, trot, and canter, the planters appended two more modes of moving: the slow gait and the rack. These are high-kneed, four-beat gaits: the rack is fast and flashy—the hooves strike in even rhythm—while the slow gait is stylish, more restrained—a shuffle in the rear, out of time. Both are hard on the horse, as each hoof hits the ground independently, carrying for a moment the horse's full weight, but they are smooth to ride. Both allow the rider to sit straight up on his cut-back saddle as the horse rushes across the fields (or, lately, around the show ring to the sound of organ music), breathing hard.

Though, by this point in the course of guided evolution, some foals rack right up off the ground, most are born three-gaited—just plain horse. To train them to their gaits, some trainers will tether a filly's back legs with rubber tubing or shackles lined with fleece; others will ask them to serpentine quickly down a slope. The horses take it up easily, usually; once they get into the habit, you'll see a five-gaited mare racking in her pasture, chasing the chickens. But if they don't, some barns resort to weighted shoes or even soring, a practice in which a bad trainer will paint the tender soles of a horse's hooves with a blistering ointment, until she flies.

See it: rider and horse spin around an indoor ring deep with dragged footing. The rider's coattail flutters, the horse's tail flags—a blond streamer that trails along the ground. The horse's eyes roll white. Her ears are pinned. Her front feet flinch in unvoiced yelps against the air.

Ask some trainers if this is cruelty, though, and they'll dismiss you. Pain is beside the point; style is everything. Once a rack is established, the gaited horse is like a car with a luxury suspension. The rider sits at speed.

When my brother was a toddler, my mother bought a five-gaited Saddlebred colt, and I named him Toby, after the basset hound in *The Great Mouse Detective*, whom he did not resemble even a little bit. Toby did, however, look a lot like Ruzon. With a dark chestnut coat and four white socks, he was the Arab's echo (unlike Ruzon's swaybacked half brother Gris Gris, my first love, whom my mother also owned). Toby was fiery, too, like his predecessor—or at least my mother read him that way. To me, he always looked a little wild-eyed, a little spooked, as if he were being asked to do more than he was entirely sure he could.

Toby, alas, was a grown woman's horse, not a girl's. He didn't get to run flat out along the levee or eat a convent school's azaleas. My mother boarded him at a training barn where he was trained for showing, his back legs tied together with tubing sometimes, his front hooves occasionally blasted by a kid wielding a fire extinguisher until his front knees went up high—*Up! Up!*—so very, very high. He had a flashy rack. Maybe he'd make it to the World's Championship in Louisville one day. He'd better, since he was a tax write-off (some lawyer's idea), a fact that would bring us all a lot of grief down the line.

To show him, my mother had one bright blue show suit and one brown, which complemented Toby's chestnut coat. The suit jackets had wide skirts, and the pants were belled to accommodate hard-shined lace-up boots. She wore this ensemble with a brushed brown derby, her red hair pulled back tight and twisted into a netted, unobtrusive bun.

This was a kind of sanctioned drag, a man's suit worn by women who wished to ride men's horses. A "lady," of course, would never ride astraddle. Until the early twentieth century, etiquette required that "gentlewomen" ride in skirts, sidesaddle, to protect their chaste reputations (and, ostensibly, the integrity of their hymens). Perched dangerously on one pelvic bone, a lady kept one foot in a stirrup, the other crossed awkwardly between two pommels. In fact, the second pommel, the leaping horn, is a relatively modern innovation that allowed the rider to safely hold the reins; in the Middle Ages, sidesaddles were so insecure that a woman riding in one was unable to control her own horse. Her mount had to be led by a walking servant or another mounted man.

As my mother is fond of telling people who challenge her equestrian bona fides, she could jump sidesaddle. She won a silver punch bowl sidesaddle. But the rest of the silver in my parents' house—trays and julep cups, Revere bowls, enameled candy dishes, sterling tea sets, a two-foot-tall Sterno-heated coffee urn—she won in men's clothes, a man's hat, beating men on their horses, on their turf.

This was the life she'd made for herself—the challenge she set, and kept raising—inside the barn and out. *Did you know that your grandmother flew planes?* I ask my daughter, every year or so. *Did you know that the dean of the law school told your grandmother she'd make a "fine legal secretary," the day she graduated at the top of her class?* As the first female maritime lawyer in Louisiana, my mother climbed down into the hulls of cargo vessels wearing stockings and heels. She stood in the doorways of the men-only clubs where conferences were held, her back against the jamb, and made prospective clients bring her drinks. Her beauty helped her go places no woman had gone before—when I was little, she looked like a redheaded Grace Kelly or, maybe, Samantha from *Bewitched*— but she needed all her strength to back that beauty up. And that strength came from training, what riding taught her how to master. She'll tell anyone who's shocked when her refined manner turns fierce, *Darling, I grew up in a barn.*

Under the lights in Germantown, Tennessee, my mother and Toby beat the trainers in the Five-Gaited Open. In the picture on the bookcase, she's in her blue suit, flushed and grinning in the flash as they hand her the silver urn. Sweat lathers Toby's shoulders. His nostrils are wide and red, blowing. His ears prick. Fear had made him high; there was a zoo down the hill from the arena, and he could smell the lions. My mother could smell them too—but she had never been happier, felt better vindicated or more alive.

Every summer, from the time I was eight until my early teens, while my brother and father traveled to baseball tournaments, my mother and I went on the road with Toby, horse-showing. Days were full of Waffle House grits and Hardee's ham biscuits, books on tape rented from Cracker Barrel, and naps in the Suburban's wayback. We spent the nights at "horse show motels" in tiny Southern towns. Their carpets were never clean. In a Harrodsburg Travelodge, my mother stole sheets off the housekeepers' cart and laid a path for us from the beds to the bathroom. In a Best Western in Paducah, she pinched a stranger's abandoned negligee from a hook in our closet and threw it into the parking lot, a slick black bird flying over the balcony rail. Even after it was gone, the smell of sex loitered in the room.

I couldn't identify the odor then—some secret musk that made my mother's face twist up—but I overheard her talk about it later with her barn friends, who often bunked with us or in the room next door. Laughing beyond the room's connecting door, they whispered *prostitute, john, strip club, lady of the night*—words I'd look up in the dictionary when we got back home.

Debra, with her satin sleep mask that reminded me of the horses' blinders, and Mrs. Fox, whose trunk was always full of Lindeman's chardonnay, would caravan with us to almost every show, parking their vans beside ours in the motel parking lots. Late at night, I'd lie under the thin blankets, reading my mother's old *Black Stallion* books, listening to the women gossip about horse sales, trainers, other people's marriages, who could *really ride* and who couldn't. Debra's laugh was a crowlike cackle. Mrs. Fox did a lot of hollering into the phone. At home, my mother was all designer suits and dessert forks and discontinued French perfume, but this, I thought—*this!*—was real adult womanhood: footlockers full of makeup and ziplocked bobby pins, emotional combat, full ice buckets, and plastic cups of wine. Womanhood smelled like sweat and hairspray, and it was loud. Sometimes, you had to yell if you wanted to be heard.

I brought a friend of my own only once, and when we snuck out of the Shoney's and up to the room during dinner to take what would be childhood's final two-kid bath, my mother and Mrs. Fox called the police, thinking we'd been abducted by a man they'd noticed at the bar. We'd drawn his attention—*the way he looked at you*, they said. This, too, was something I didn't understand for a good long time.

Going to the show barn in the morning was a pleasure. Clutching my Styrofoam cup of grits, I'd stand for a moment in the door to the waking barn, inhaling the smell of fresh shavings and liniment—that sharp, mentholated clean. Everything was in order: the tack room lined in colored silks and hemmed by potted hedges, the feed room full of neat cubes of plastic-wrapped shavings and bales of hay. Outside, in the arena, the trainers warmed up horses—big men hunched, whips up, over the backs of hobbled, glossy things.

During the day, I was alone while my mother trained and watched the barn's other riders. I hot-walked pleasure horses, poked handfuls of grass through the bars of the raffle ponies' pens, hunted for four-leaf clovers, got electrocuted (mildly) by the popcorn stand. But, mostly, I sat on restaurant-size bags of carrots and watched the grooms set tails.

Saddlebreds' tails are lifted high and grown out long so that, in the show ring, they trail like pennants, an extra indicator of the horse's speed. To achieve this look, a horse's tail is set—the tendons progressively stretched or surgically nicked until the dock can be lifted upright, then held up vertically from the rump by a harnessed leather bustle. Resetting a tail after a class could take a groom the better part of an hour, first to pick the tail, strand by strand, free of any shavings, then to tie it, buckle the harness, wrap the crupper, arrange the feathers, tie down the dock, net it. Then the tail bag went on and was tied to the blanket.

I never learned to do this; my job was to keep the peppermints coming. (*Hold your hand flat, tuck that thumb in. He's being careful, but you don't want him to chomp the wrong thing.*) I was supposed to make as much crinkle with the wrappers as I could, to distract the horses from their other end. Left unoccupied, the horses danced in the cross ties as their tails were set—ears back, nervous, anticipating a night of discomfort, their docks rubbed bald, the flies un-swished. It did not once occur to me that this was cruelty; what I saw was care.

It was no different, really, from the way my mother got ready for the show ring, painting her face with bright pink blush and too much lipstick, curling her eyelashes, driving bobby pins against her skull. This was all extra—waterproof stage makeup that she put on thick and took off later with gobs of Pond's cold cream—but even when she wasn't wearing boots, her high heels pinched her toes, and her stockings left red lines around her tiny waist. Beauty came at a cost— that much was clear—but you had to be beautiful to win.

After working, the horses would be hot-walked up and down in the grass between the show barns until they'd stopped blowing. Later, cross-tied in the aisles, they stood sultry to be rubbed down with linens, their slender, shining bodies steaming as a groom mopped grease on their steeply shod hooves or sponged liniment over their legs, up under and beneath their bundled tails. Most of the horses were geldings—castrated to relieve them of their wildness, that selfish, bite-prone edge— but it was hard to see that. They all looked like women to me.

When I was eleven or twelve, the Equitation Champion and her horse arrived at the barn from Kentucky, and, soon after that, she moved in.

My parents were always "adopting" people and animals into our safe and comfortable home—stray dogs and stranded exchange students, fainting goats and feral cats, a friend recovering from a fall, itinerant a capella singers, the contractor's raccoon. From the ether, I gathered that the Equitation Champion's father was "nasty." I had no idea what that meant. All I knew was that she felt safe with us, not in the dorm. She felt safe at the barn, but not alone.

I related to that feeling. The barn was not a refuge but a place where you went to learn your defenses—how to protect yourself around large animals and lewd men. It was a place where the hazards were obvious and manageable (*Touch the horse's rump when you walk behind him. Keep your heels down!*), a place where you made yourself tough. (*Watch the farrier work—maybe you'll learn something. Just kick him if he puts his hands where they don't belong.*) My mother was certainly tough. She'd hit a man once with a punch bowl when he tried to grope her, bashed him on the top of the head *just like he was a nasty little stud.* The lesson seemed obvious: if you could control a horse, you could control anything.

Into our house the Equitation Champion brought the things that had helped her take control: Molly McButter in its blue sprinkler can, SnackWells, SlimFast, potato chips fried in olestra, the fake fat that went right through you. Out on the show circuit with her, the Cracker Barrel menu became a freak show of fatty slabs of beef, seductive cobblers. My mother, the Equitation Champion, and I took to eating unbuttered grits and dry potatoes. We no longer stopped at Dairy Queen. Our Wendy's salads went undressed.

To control a horse's body, you must first control your own. *Heels down! Chin up! Hands still!* Imagine an iron poker taped along your spine. Imagine a rod running through the center of you, from the top of your head to the backs of your heels. *Up, down. Up, down.*

The Equitation Champion posted like a piston, converting the horse's horizontal flow to her vertical. *Up, down.* She smiled steadily around the arena, her face unmoving even as her body kissed the saddle and rose, kissed the saddle and rose. Nothing jiggled, nothing bounced—she was that thin. I leaned on the sticky creosote rails of the ring and watched, as I'd been told to do. Out in the field, Gris Gris, Ruzon's half brother, swished his tail, grazing. I wanted to be out there, too, lying in the clover and watching the clouds, or hacking bareback through the pines, or down the road at the drive-in, slurping a strawberry milkshake, but if I wanted to ride—*really ride*—I needed to see how it was done. This kind of control was what my mother wanted for me. I didn't have it.

My body, thrashing through puberty, needed mastering. I was no Equitation Champion, not in sufficient control of my body to control a horse to the trainer's satisfaction, not to be trusted with easy, unbroken speed. I rode in Three Gaited Country Pleasure, wearing a yellow-pinstriped brown hand-me-down suit, so close in color to the bay coat of my leased gelding, it looked off. Showing Spirit, I couldn't keep to the corners, never remembered to check my leads.

After an early morning class in Jackson, my mother waited outside the ring, swatting her boot with a white-handled whip.

You were out in la la land, she said, exasperated.

Hormones, I know now, were responsible. My brain fog would vanish, along with my period, after I'd become too thin.

Once, though, I did manage to come in second place. In the photograph, a red ribbon is hooked to Spirit's headstall, and my red face shines, spotlit. My chin is tucked into my second chin while Spirit's tucked chin is creased, his jawbone sharp as a blade. But, even in my triumph, something is not quite right. My ears—elfish things that stick straight out—are invisible in the picture. My mother had superglued them to my head.

In my thirteenth year, my mother and the trainer suffered a rift. Something about the farrier, her opinions about Toby's shoes. Something about how he didn't like it when he got beat. When she tells me the story now, she sighs—stale reproach—and says what she wouldn't have said then: he didn't like that she was a woman, and that she knew more than him.

It all happened fast—my memory's a blur. She pulled Toby out of the barn, moved him to a top-notch training barn in Indiana, where, on a visit, the Equitation Champion ran miles through the pastures in the snow. When we got home, I switched to hunters.

At the new barn, there were no tail sets, no derbies. Instead, we ripped the horses' manes with a pulling comb to make them easier to braid. We still wore drag, but now it came in the form of spandex britches and high, calf-hugging boots, stock collars and jackets in somber hues. At the new barn, the vet was always there, arm-deep in a Warmblood, palpating ovaries or untwisting a gut.

My period started in the bathroom of the trailer that served as the barn office. Two years later, it stopped. I weighed 130 pounds at twelve; at fourteen, I weighed 106. I accomplished this by eating a diet a girl might feed her Breyer horse: one half of a bale of shredded wheat for breakfast, one apple for lunch. While my parents worked long hours, I took over the grocery shopping and cooking for my family, "perfecting" our family recipes until they contained less than eight grams of fat. Every night, I did six hundred sit-ups in front of the TV.

I write "accomplished" because the control I had taken over my body was treated absolutely as an accomplishment, rewarded as one. At dinner parties, my normally mild-mannered father would invite our guests to punch my "rock-hard abs," and my mother bought me a horse.

Older than they told us he was, Nate was sweet and slow, big and almost black. He arrived at our barn in mourning. Unsure of where he was, missing his former rider—a girl named Kirsten, who I imagined in braids like the American Girl doll of the same name—Nate wouldn't eat, and so I stood in his stall for hours every afternoon, feeding him by hand. He lost two hundred pounds.

Still, we got straight to work, learning how to hew to corners, to see distances, count strides. When the vet found us in the ring alone, he'd mount his stocky little jumper and give me lessons without my consent. He'd tell me to drop my stirrups at the trot, keep posting. *Up, down. Up, down*, he'd yell sarcastically, as if it were a joke. I'd post until my calves were raw. Until they bled. I had no choice. If I stopped posting, or picked up my stirrups, the vet and his horse would crash into us, broadside, driving us into the rail, so that my knees slammed into the fence posts at speed. Sometimes he'd use his crop to keep Nate moving. Sometimes he'd take my reins so that there was nothing Nate or I could do to get away.

One night, during a lesson, the ring slick and steaming after a summer rain, I pushed Nate, my legs trembling, into the corner. When the vet yelled, *Switch leads!* I pressed a spur into Nate's side and squeezed the reins. His legs went out from under him. He fell. I fell. His body lay on top of mine, until, dazed, he scrambled upright again, knees shaking, and crushed my right hand.

All this happened so fast, I didn't know I'd been hurt until Nate hung his head in a gesture of contrition, his eyes hooded, and began to blow. Pressed into the wet footing, my hand was bloody and already swollen. Though no bones were broken, it would remain the size of a softball for a month, a gauze-wrapped bundle I carried around with me like some sort of prize. For a long time, I showed it to everyone I met; it was my best scar, a badge of toughness—this smooth white hoofprint across the freckled bones of my hand.

Eventually, I figured out that if I punished myself enough, no one else could do it for me. I worked harder, ate less. The summers I spent living in Nashville to train at a better barn, I ate a sweet potato and a box of frozen spinach for dinner, every night. After I'd remained at 106 pounds for a whole year, my mother and I flew to New York, where I was fitted for bespoke knee-high boots to wear at my first big hunt show. Everything about the shop was beautiful. I can still feel the smooth solidity of the wooden lasts, smell the leather: baby calf, French calf, oxblood, pony skin, ostrich. I shivered as the cobbler bent to run a tape around the arch of my foot, up my shin.

So small. Elegant, he said, nodding, as he measured the muscle of my lower leg.

The boots, for a long time, were my most prized possession. So expensive. So black. While my mother picked her horse's hooves, I sat on the floor of the tack room and polished them to a high, hard shine. The boots were rigor itself: mastery, control. They had no zipper, no snap, no gusset, only a short, waxed lace that crisscrossed over my foot's high bridge. To get them on, I slipped two wood-handled hooks into the loops sewn into the boots' lining and pulled. They fit like tougher skin.

I was a teenager, though, and I was growing. The boots did not grow with me, though I maintained my 106 pounds. I checked in on my inches by putting on the jeans I'd bought when I first hit that number—a white pair from The Limited with no stretch. Occasionally, I'd slip a little—eat, for instance, Thanksgiving dinner—but my body, by that point, was so unused to fat that butter made me sick. Nevertheless, all that posting without stirrups, all that running—my muscles grew. Before each show day, I sat on a tack trunk, tugging. I used silicone spray, baby powder. I wore knee-high compression hose over my taped-down britches. Eventually, a groom would have to sit on me to keep me from sliding, while my mother and my trainer pulled, one on each hook. Blushing hard, I'd let them wrench me into the boots then run away to the safety of Nate's temporary stall.

At the end of all this, Nate and I looked great—*Nice*

turnout!—his coat curried and brushed to the same high black shine as my boots, my hair in a close-pinned bun, his mane ripped down and braided tight. My fingers were not strong enough to do it well, and so we hired a braider who'd come in at dawn the day of a show and stand on a step stool, wrapping the coarse hair around her chapped fingers, stitching each tiny braid through itself with a fat silver needle. Nate hated every second of it. If we didn't Velcro a spandex neck wrap around him, he'd rub the braids loose in an hour against the door of his stall.

I was eighteen when I rode in my last horse show. In the photograph, I smile next to Nate in his tight braids, holding his reins in one hand. My cheeks are gaunt, my collar gaps. My belt's tail vanishes behind me. The boots are beneath the frame, but I can feel their tight grip on my calves, and I know they shine.

I had achieved it, finally: a blue ribbon, womanhood. A year after that picture was taken, I was presented as a debutante at a Carnival ball, wearing a plume like a carriage pony's feather in my hair. My father squeezed my hand in its kid glove, laughing, *Isn't this just like a yearling sale?*

Competitive, and good at learning lessons, I got into the college I wanted, where I yelled to be heard and drank plastic cups of wine. I never hit anyone with a punch bowl, though. Nasty little studs didn't bother me—I was tough. *Do you want to see my hoof-shaped scar?*

Then, at the beginning of winter break my sophomore year, my mother flew up to Connecticut and picked me up in a rental car, headed farther north. We were on a rescue mission to find Toby, whom my parents had sold a few years before, that tax write-off business having finally come due. My mother had heard through the grapevine that he was in bad shape. Though the original buyer had agreed to give her the opportunity to buy him back when the time came, they had not. He had gone from a good barn to someplace not so good, and, now, lamed, he had been cast off. A Good Samaritan had taken him in and was giving him back to my mother, if she wanted him. Of course, she wanted him. She loved him—still loved him—had loved him this whole time.

We drove up the freezer-burned highways, along the coast of Connecticut, up through Massachusetts, and into New Hampshire, my mother erupting, every so often, in brief outbursts of rage: *Some people shouldn't be allowed to have animals! You just don't do that to a horse! They should be shot!* Finally, late at night, Toby's caretaker led us into a dimly lit barn. I don't remember how we found him—shod or barefoot, blanketed or unclipped, cross-tied or in a tidy stall—only that when he saw my mother he lowered his head, and she wrapped her arms around his neck, and she cried. We were all crying, I'm sure of it, though a horse's tears are invisible, their sorrow a thing you need to be willing to see to know.

Still, it wasn't until after college, when I took a summer job as a wrangler at a dude ranch in Colorado, that I finally realized all I'd gotten wrong. It took a little doing, even then.

When I arrived at the ranch, the other wranglers were just as suspicious of me—citified in my home-starched shirts—as

I was of them. I fought with my boss about wearing stiff, high-waisted Wranglers in the heat. I fought with the other wranglers about how they left the string's ears unclipped and their fetlocks long—*as God intended!*—to chase away the flies. Not trusting that I could really ride, they pulled the dullest colts for me my first week on the job—sleepy six-year-old roans that had to be spurred to a trot—and sent me to lead children's rides in the low pastures, where there was no chance I might get lost.

After that, to prove myself, I rode nothing but three-year-olds and troublesome mounts from the wranglers' herd. High-headed Okie, who had some Saddlebred in him. Pecos, who liked to buck without warning, and who, at the end of the summer, I took home. Eventually even Jeremy, a cowboy who spent his Sundays riding bulls, would watch me settle a spooked colt with an approving nod.

None of the horses at the ranch were that difficult to ride—they were working horses, after all, bred for steadiness, expertly trained. On Thursdays, in the big covered arena at the bottom of the mountain—a place the ranch's guests were not allowed—a cowboy from a nearby ranch started two-year-olds. This process, as I'd known it, was always rough and dangerous: colts rearing under whip-wielding riders, fillies overheating in the round pen. But the way this cowboy did it was so calm, it was almost silent.

One Thursday, my own work done, I snuck in through the side door and watched, hardly breathing, from the bleachers, as the cowboy circled a gray colt, slowly, touching him here and there, speaking in low whispers. Eventually, he asked me if I'd like to learn.

As I held a colt close at his head, the cowboy dusted him with blankets. If the colt skittered, the cowboy held the blanket firm against his body, but if the colt held still, the blanket went away.

It's about telling him how to act when he's afraid, the cowboy said. *You apply, and then release the pressure. Apply—*

He stood up in one stirrup, talking softly, until the colt went still.

—and then release. You teach him—

He set his weight down in the saddle, waited until the colt stopped tossing his head, dismounted.

—that no harm will come to him. That he's free and strong and powerful, but that you are not a wolf. There's no reason, he said, stroking the colt's soft muzzle, *just no reason at all to be afraid.*

And maybe this was my trouble, too. I was afraid all the time and called it toughness. I needed to learn to trust myself instead.

Three weeks into that summer at the ranch, I saddled the big mustang and took an experienced group for a daylong ride up the mountain and into the national forest. It was a trip I'd heard described but never taken, and so I asked around for a map to the valley that was our destination. There was no such thing. Riding out, I was sure we'd wind up stranded on some logging road as night fell, that we'd never find our way home. But, as I marked each turn we took—counting false trails and memorizing aspens—the forest became legible. I could almost feel my brain developing, as I engaged, for the first time, my innate human ability to navigate a wilderness. Maps were not required; all that was necessary was to rely on our instincts, trust the horses. We gave them their heads as they leapt over streams, rose from their backs as they worked their way carefully up rock faces littered with scree. Around noon, we found ourselves in a meadow ringed by mountains, dismounted, pulled our picnic lunches from our saddlebags, and lay down in the high grass. I ate that day like I'd never eaten in my life.

After that summer, I wore my high boots one last time, at a fox hunt in Tennessee. I still don't know how I got them on, but one of the whippers-in complimented my turnout while I waited on my borrowed horse for the hounds to be cast.

Riding with the second flight up the scrubby fields and through the forests of ash and oak, I was careful to follow the hunt's strict etiquette—always staying behind the field master and the hunt members, turning my horse's head toward the hounds as they passed, so that he wouldn't kick. (His name was Bucky, and it suited him.) *Ware hole!* we shouted, one after another, as the flight sidestepped a deep pool of water in the path then took off at a slow gallop. At the bottom of the forest path, I followed my mother and her borrowed gray over a coop into a little clearing. Ahead of us, the field master held his red arm high. We halted hard.

Then, off in the brush, we heard the hounds open.

Trailing the baying of the hounds through the thicket, I rode wildly. Branches swatted at my jacket's arms, clattered against the felted plastic of my hunt cap. The flight dispersed among the trees, and for a moment, I was lost—the forest seemed to close around me, and my heart beat against my ribs. Nearby, the hounds were at some creature's throat.

I urged my horse on to the edge of the trees, where the land dropped steeply to a rocky creek. Down below, the water ran red. Upstream I saw the field master's scarlet coat, a thatch of gray fur, the roiling bodies of the hounds. Then, up the bank, a rider jolted toward me at a lope, his free hand held away from his snow-white saddle pad. Halting beside me, nose to tail, he ran that hand, wet, across my cheek.

Coyote, he said. *You're blooded now, girl.*

I could taste the iron, a sickening smear, at the corner of my mouth.

Later that day, before we'd made it back to our starting point—to the hunt tea, the stirrup cups, the trays of dainty canapés—my legs began to swell in the tight boots, their circulation cut. They throbbed, a strobe of numbness and pain. At the top of a hill, I halted, took a drink from another rider's flask, and borrowed his knife. Crossing one leg over the pommel, I cut the lovely, hand-sewn stitches, set my body free.

I tell my friends with children that it's good for a girl to grow up around horses; I think it's true. It teaches you about power—how to wield it responsibly, when you must let go. It teaches you self-possession and the ability to stay calm when you're afraid. It makes you strong, I say, by which I mean: *Chin up. Shoulders back.* I don't encourage them to let their children hunt or show.

The fluffy mini pony my daughter rides gallops away each time we go to catch her. Rescued by an old friend of my mother's who runs a sort of foundling farm, Licorice was clearly hurt by her old owners, who kept her tethered in their yard.

At four feet and ten hands tall, respectively, my daughter and the pony see eye to eye. Alone with her, the pony knows she won't be overpowered. Her ears stay up, her eyes dark, her head low, as my daughter moves around her with the curry-comb and the brush. Her coat hardly gets any cleaner, but I don't interfere. I stand outside the stall and listen to them talking. My daughter, soothing, murmurs, *Oh, sweetie sweetie. Oh, poor you. What happened?* The pony, apparently, gives her answers. *Oh, no, they couldn't! They did? Why would a person want to hurt a little pony like you?*

Still, even a year into this gentling, catching Licorice in her pasture is a process. We go out whooping, swishing sweet feed in buckets. Then, as the herd comes running, we leave the buckets on the ground inside a smaller paddock, leave its gate open, go and hide: a trick. While the pony's head is down, engaged in eating, I skulk along the paddock fence like a wolf, shut the gate. Inevitably, she startles as I approach, softly saying *whoa*. The slightest screwup—rope over my shoulder, a step too fast—and she'll go running, clever in the way she slips my grasp, fast as all get-out. Before I can even stand up from my crouch, she'll be halfway across the long green meadow, her little legs churning swiftly beneath her, her body doing just exactly what she needs it to do.

WHAT WILL LEAVE YOU
ADRIENNE CELT

When I turned thirty, I decided to take horseback-riding lessons, in part to research a novel I was writing, but also to appeal to the child inside me, who'd always longed to be a cowboy. The time was finally right. Every week, I drove to a dude ranch near my home in Tucson, winding twenty-five minutes through Saguaro National Park and watching the desert cycle through its seasons of dryness, greenness, and glorious bloom.

Sometimes in the winter I'd arrive after a rain to find the horses drenched and the conditions less than ideal for a lesson. This was its own sort of education: I didn't know that horses could turn their ankles in deep mud and go lame. I didn't know how fragile those large, elegant bodies were.

In fact I didn't know anything: how to pick a horse's feet; how to cinch a saddle in just the right place, with the right degree of tightness; how to slip a halter over a horse's long face. Often they would elude me when I tried to do this out in the pasture, flicking their ears and then trotting casually just out of reach. I learned to whistle at the ranch dog, a sweet and enthusiastic Australian heeler, and then point her at whichever horse I wanted to ride. Sometimes, we had to do this four or five times before I could get the halter and lead rope in

place, and I'd get frustrated and embarrassed as I walked behind the dog, ropes dangling from my impotent fingers. But
all this was useful—my ignorance was the point. The novel
I wanted to write was, in part, about learning horsemanship
from the ground up, in a sudden blaze, so it made sense for me
to start from nothing.

There was a scene in my book, very much informed by those
lessons, in which a young man exercised an energetic young
horse in a round pen, urging it into faster and faster circles
with the crack of a whip, and then turning it around to run the
other way. It was a moment of casual expertise. As he stood in
the center of the pen, dust rose up around the rim in uneven
intervals under the horse's gait, lifting into the sky like the
points on a crown. The moment still comes to me sometimes,
even though it's been years since I first wrote it. Like the scent
of hay being forked out to hungry mouths in the stables, and
the feeling of moving into a canter for the first time—sudden
smooth undulations, following a choppy trot—it is embedded
in my brain.

A few months into my lessons, the owner of the ranch fos-
tered a mustang that had been badly burned when (according
to local rumor) a cartel had torched the barn of an informant
on the US side of the border. The horse was too spooked to
ride at first, so the owner tossed him into a pen to let him run.
Eventually, he reacquainted the horse—who he'd lovingly, if
dickishly, named Bernie—with being saddled, then ridden.
I watched him, week after week, stoically maintain his seat
as Bernie tried to throw him off, bucking and snorting. And

I saw the moment when Bernie changed his mind. It was as simple as that: he looked at the man, who was slipping a bit gently into his mouth, and decided he was no longer a threat. After that, Bernie grew to love humans, and would literally eat from the palm of your hand.

I also think about my own horse, Lady, and all the hours I spent exercising her in a round pen before riding, watching her lag petulantly in the beginning, and then find her spirit and kick into a higher gear. The physical memory of the lunge line in my hands, a bright blue rope that clipped to her halter. The way she snorted when she really got going and pumped her legs like pistons.

But I'm getting ahead of myself.

I wanted to write a Western, but more than that, I wanted to be in a Western: to disappear into the desert with my trusty steed and learn the sweat and pain of a new lifestyle; to be incompetent for a while, so the mastery would be all the sweeter when it came. I wanted to escape the world I'd built for myself, which focused so much around my computer, around being inside and reading small words on a screen or page, scanning them for some kind of cathartic energy. There must, I thought, be other methods of catharsis.

Even though learning to ride a horse had been a dream of mine since childhood, I had to trick myself into it, wedding the lessons to something productive: research for my book. I tried to use everything I experienced. My heart, pumping with unfamiliar joy as I raced down a dirt road in the desert, became my protagonist's heart. The smell of creosote thick in the air around me, and the small Quarter Horse moving into a canter with the merest squeeze of my legs—these were given to her as well. She was not me, but we had things in common. We both lived with deep nervousness about the future, our families, the possibilities the world might offer us, and we both found solace on horseback. The difference was, I chose my new equestrian way of life, while she was thrust into it. This—the protagonist's lack of agency—would become a recurrent problem with the manuscript, something I would struggle with for years. But I didn't know that then, in the early drafts. Her hostage energy made sense to me. Her sense of escape, once she grew to embrace her new life under the desert sky, reflected my own.

But it wasn't only escape I sought: it was connection. Epiphany, even. As I learned, week after dusty week, horsemanship is a conversation, a language you and your horse co-create. You can teach an animal voice commands, or you can make your requests silent and soft: tilting forward or back, movements imperceptible to the eye. You, the rider, must learn to listen to them, too. What it means for your horse's ears to flatten in fear and anger, or perk upward in sudden alert. The texture of a snort, the licking and chewing of comprehension. Riding well is a flow state not unlike writing, a place where you exist outside time.

Whenever I pulled into the dirt parking lot and waved hello to the ranch's owner, this new creativity blossomed in me, along with the equally important knowledge that I had chosen to find it. Riding, like writing novels, filled me with the conviction that I was lucky, that I was strong, that I could do most anything I wanted to, if I was simply willing to try.

I met Lady—a paint mare with a lot of attitude and a stocky build—at the dude ranch where I took lessons. But she didn't officially become my horse until a couple of years later, when I made the impulsive decision to buy her. It all happened quickly: one Friday morning I showed up for my lesson, and there was a stranger picking out horses to bring to a different, more fiscally stable ranch in New Mexico. A barn hand told me that our place was closing. Bernie was already gone, as was Jackpot, a filly I'd been watching fatten up ever since she was born. The rest, including Lady, would go at auction, which, given the inherent randomness and lack of regulation at horse auctions, could easily mean going to her death.

Standing in the dirt pasture, my hand on Lady's neck, I called a horse-owning friend of mine in tears and explained the situation to her. If I bought Lady, I could save her. "Are you sure you're ready to own a horse?" my friend asked, gently.

I was not sure. But what else could I do? Abandoning Lady to her fate did not seem like a feasible option. I didn't know what owning a horse might entail, but I knew *her*: Lady, with her broad forehead and flicking tail. The absolute line between the white in her mane and the black. How she would roll around in the dirt after I gave her a bath, falling over one way and then the other with all her immense weight, and then jumping up to shake off like a dog. At that point, I'd been riding Lady exclusively for months, and she and I had formed a bond. No longer did she race away from me when I went out to pasture to bring her in for a lesson.

With my friend's help, over a frantic few days and innumerable phone calls, I found a nearby stable that had space where I could board Lady, and brought her there with all her tack, which I heaved into a plywood locker. As my friend and the barn owner stood chatting, I felt dizzy with responsibility. They, the lifelong horse owners, knew so much that I did not. I was committed now—my relationship with riding had suddenly bloomed into something much bigger than research for a novel. A horse can live over thirty years, and Lady was only nine. Somehow, the conversation turned to burying dead horses with backhoes, and this was the thing that broke me. *Can I afford to rent a backhoe?* I wondered desperately. *Do I want to rearrange my whole life around a creature I'd need to rent a backhoe for?*

Maybe what I meant was: *Am I capable of doing this at all?* It seemed portentous that, on my first day as a horse owner, I was being called upon to contemplate the proportions of her grave. Perhaps, I thought, the universe was trying to tell me something. Teach me a lesson about overreaching.

Lady did not die. I did not, despite my inexperience, ever bring her to any harm. But when I look back on that conversation now, it still feels like a piece of foreshadowing. The way that endings are implied in every beginning, as much in life as in a story. How choices, even when they're yours, can still feel like they're beyond your control.

I did get used to being a horse owner. I drove to the barn, forty minutes each way, three or four times a week, even though that first summer I was also working two jobs, revising my novel, trying to have a life. I was busy and tired, but being with Lady was like letting go of a deep and painful breath.

She was tricky to ride at the trot, bouncy and easily distracted. A bossy mare, but bad at leading on trails. She liked having her way, but not taking charge. There was a difference. Very quickly, I began to project my own qualities onto her, as people do with the animals in their lives. *We both like James Taylor*, I thought, when I sang to her while warming up. *We're both try-hard, we both get frustrated, we both feel bad when we have cramps.* With the help of a trainer from our new barn, I softened Lady's trot until it was smooth and pleasant. With patience, I learned to read the signs that indicated she was in heat, or had just eaten, or might be injured in some small way. My trainer and I were very different people—she was older, conservative, and a bit of a daredevil. She had once competed in Three-Day Eventing, an exquisitely dangerous type of riding that involves galloping cross-country and jumping over solid objects. But we talked during lessons and found we were both sarcastic and cynical, and enjoyed making fun of the men, full of bluster, who came out to ride the barn's obstacle course, only to be unseated within minutes.

"You're good with her," the trainer told me, nodding to Lady. "And she's not a bitch, like some mares I know." After the election in 2016, when I was feeling like a bitch myself, I went out to the barn, ignoring the Trump flags I saw flying in the neighborhood nearby, the "Killary" bumper stickers. My trainer, a likely Trump voter herself, was wary around me for a while. "I'm not racist," she told me one day, in the middle of a story about a friend of hers playing loud mariachi music. "Okay," I said, not knowing what else to say. Lady was the same as ever. She was the still point in my world, while politics and emotions spun around her. She and I stood together that hot November, while I pressed the bit gently into her mouth and pulled the bridle over her ears. I adjusted the girth of the saddle, ran my fingers through her mane, and lifted myself onto her back. There was no wind, there was no sound, except the birds overhead, the generator humming into life. Together, we survived.

Meanwhile, the novel I was writing began to slip away from me. I remembered the excitement I'd felt the night I wrote the first few pages, how I'd read them over to myself in bed and my pleasure glowed within me. *This is something real*, I'd thought. *I will make my name with this*. It was my second book, meant to telegraph my range to a wide new readership, a marker of my ability to write about violence, tenderness, wilderness, suspense.

Now, when I reread passages for revision, all I felt was exhaustion. This was the book that had led me to Lady in the first place, but spending time with her didn't help clarify the issues I was having with the novel's plot. The horsemanship was working: it was the human characters that gave me trouble. I loved them all too much, it turned out, to make them act realistically. I wanted to save them, both from their worst impulses and from the world. But doing so meant I was no closer to publishing the novel and reaching my imagined readers than I had been several drafts before.

How could it be, I wondered, *that my best intentions are making everything turn out so badly?* Given that it wasn't my first novel, I thought that I knew how to write them by now. That getting over the hurdle of my debut, a few years before, meant I could take any story I liked and turn it into a part of my career. I didn't realize that books are fragile. That you can crush them; that they can crumple beneath the weight of your expectations. That they often remain private and un-published, a period of your career that is invisible to every-one but you.

One September, about two years after buying Lady, I left for an artist's residency in Sheridan, Wyoming, trusting my horse to the steady hands of the barn staff, and taking a step into the unknown. As I was packing for the trip, I had to make a decision: to bring along the pages of my long-suffering novel and start yet another revision, or to leave them behind and use the unstructured time to begin something new.

I'd been working on the novel, on and off, for years by that time. Lady had found her way into the book under the guise of a fictional horse named Candy, who was a gelding with different coloring, but her same rugged body and picky mannerisms. And yet, the two things—my riding life, and the novel that had presaged it—didn't feel as inextricably linked as they had at the beginning. The connections were more oblique. Was Candy also Lady? Yes and no. It was possible for me to imagine moving on to a new story, and still riding my flesh-and-blood horse. In fact, I'm sure it would've seemed silly to me, then, to talk about them in the same breath.

The residency would afford me a month away from my day job and other responsibilities, and I knew I needed to take advantage of this rare opportunity to justify leaving everything else in my life behind for the sake of art. The old book haunted me, but the contradictions I'd long felt, between the characters and the mechanics of the plot, were still there, impossible to reconcile.

When I stepped onto the plane, I brought only my empty notebooks, fresh pens, the blank page.

That September, the air in Sheridan was yellow and hazy, as smoke from the California wildfires drifted over and settled in. It changed the quality of light, and so, even though it was unhealthy, I didn't mind it. I would leave my studio and walk down the dirt road, over the river, past herds of deer. The deer would startle when they saw me, but then pacify. All I ever did was lean on the fence posts and watch them, from a distance. Sometimes, when I walked back home, the residency's white cat would vogue for me on the driveway. I wrote half of a new book in those weeks.

Every night, the other residents and I would gather around a big table and eat the food one of us had cooked. We talked about the people we'd left back home, the projects we were pursuing, the things we loved. One man, who'd raised horses in his youth, was creaky with age and from the years he'd spent being thrown, jostled, and stomped on by his charges. Still, he told me, "Hold on to your girl, if you can." The pressure of this directive made me uncomfortable. And yet, I assured him that I would hold on to Lady with all my might, because, at the time, it felt true.

I could not yet put into words the new sensation I had, of drifting. The way the tentacles of responsibility had loosened from my neck, my arms, my tight-squeezed lungs, now that I was far enough away from my horse that her beauty and her personality were not there to compensate. Gone were the long drives to the barn, the hours spent on tedious, if loving, chores. Even before leaving for Wyoming, I'd begun to doubt whether I could really afford a horse in the long term; always there were unseen costs, from a new saddle pad to an urgent vet's visit. (*And someday*, I still knew, *the backhoe*.) Now, my energy was pouring into my new novel, and I didn't want to slow down.

But I couldn't break my own heart twice, in the space of such a short time. I'd left the novel. I told myself I'd keep the horse. It was a way of holding on to the person I thought I was, even as I became something else entirely.

Around the halfway point in the residency, we all got stir-crazy and decided to venture out to an event called Don King Days, which celebrates a famous saddle maker from the area—not, you know, the other Don King—and showcases a number of equestrian events, mixing the ruggedness of Western apparel with the unmistakable whiff of money. The fairgrounds were full of sno-cone machines and barbecue stands, screaming children and laughing adults; there was a tent full of small molten forges and flying sparks, where men and women built like Mack Trucks competed in the World Champion Blacksmiths Contest hammering iron into free-form horseshoes. Polo was being played, politely, on a field nearby, the spectators sipping free mimosas.

I remember that day for many reasons. Before the Pledge of Allegiance, two men jumped out of a helicopter and crisscrossed in the air while holding a large American flag between them. Apparently, one of these men had lost both his legs performing the same maneuver in the Army, during a jump that also, as it happens, killed his fellow paratrooper. Bionic of leg, now, he had re-enlisted in the military, and was here today to celebrate America. He landed safely in front of us, the flag floating gently beside him. The cheering was loud; the mood, surreal.

After the parachute jump, they set up for calf roping on the same, slippery, fenceless field where the polo match had been played. Usually, rodeos are run in dirt arenas, where the calves and ropers alike can find purchase, and the change of venue was nearly catastrophic. Several times, the little cows— "little" meaning up to 280 pounds, with prominent horns— slipped away from the cowboys at the last second and barreled straight into the crowd. I tried to capture it on video, but most of my shots were ruined as I scrambled away from the action, fear of death trumping my desire to record. We left when the wranglers brought out bucking broncos, because the broncs, too—lacking horns, but almost twice as large as the cows—ran right at us, furious at having been tied up by their balls. There was a cheerful sense of danger, everywhere. A sense that no part of the proceedings could possibly be insured.

It was also the kind of event where half the crowd was milling around on horseback, and by the time we left, I remember feeling distinctly less-than, left out, missing Lady. I took a last, mournful look behind me, at the cowboys galloping across the glimmering fields, the blacksmiths in their tent, hammering away like the gods. *I could belong here*, I thought. I had the expertise by then. I knew this way of life like I knew the back of my hand.

But a strange thing happened as we pulled away from the parking lot—and this I remember most of all. As we drove, my mind slipped back to my studio, and the work that was waiting for me there. Thoughts of Lady drifted away. It made me feel guilty at first, but the guilt didn't last. For the last couple of weeks of the residency, I barely thought about her at all.

I could talk forever about the smallest moments of my time with Lady. The ways she changed my life. The cowboy hat I bought at an antiques store and had to wear home on a six-hour plane ride; the now-intuitive awareness of temperature changes and how they can cause too much wind; the feeling of her gait picking up to a canter, and the two of us turning through a clean figure eight. What's harder to say is this: a year or so after returning from the residency in Wyoming, I decided to sell her. I wanted to start writing full time, and I knew I wouldn't be able to do so while also giving Lady the care and attention that she deserved. So I found a nice woman who owned property with trail access and private stables, and together we agreed on a price. I don't regret this decision, exactly: it got me exactly what I'd hoped. But still. You can imagine. She was not just a part of my life, she was a part of my identity. Not just the kind of person I was, but *the person I was*. Until, I guess, I decided I wasn't.

The last time I rode Lady, I didn't know it was the last time. I'd been planning to come back to the barn the next day to say goodbye, but my trainer, who'd helped me set up the sale, suggested at the last minute that it would be better if I did not. She wanted to keep Lady healthy, since the buyer's vet was coming out soon to do a final assessment, and with horses, you never know. They can turn an ankle, they can get pricked with a cactus, they can swallow too much air. It was good advice, but still a shock, which I felt detonate, sickly, in my stomach. I took my time grooming Lady after my ride, and then turned her out in her stall, to roll the clean off of herself, like always. I took some pictures. Then I walked away.

Before putting my old novel aside, I revised and rewrote it one last time. I typed the whole thing out again from scratch, and when I still could not fix it to my satisfaction, at least I knew I had done all I possibly could. Leaving Lady was not the same. As I drove away from the barn, I sobbed until I thought I'd break in half, saying out loud, to no one, *I don't want to, I don't want to.*

But I did want to. I must have.

I used to sometimes go to riding lessons with a very mild hangover, just a sleepy one, not a headache one, god forbid a vomit one, and those lessons always went very well. My natural inclination toward anxious perfectionism was smoothed away, and as a result my posture softened. I telegraphed my turns more elegantly, and read my horse's body language instinctively, especially if I was riding bareback and could feel her breathe: warm, like an enormous, animate teddy bear. I would breathe deeply, then, along with her. Let myself bend, fold, settle, ache. I try to re-create that feeling, now, without her. Not so much "be hungover" as "be gentle." Let yourself, occasionally, relax.

When I finally sold Lady, I had to meet my trainer in the parking lot of a strip mall to get the envelope of money, meeting in the middle between our respective places of work. It felt like a drug deal, and I left with the sensation of grime coating my fingers and my heart. It took me a long time to get over the suspicion that I'd betrayed Lady by selling her, even though I did so in order to improve her life as much as my own. Maybe I'm still not over it. Friends remind me every so often that it was the compassionate decision; I sent her to a plot of land where she had acres to graze, a new and loving companion, unhindered by the creative ambition that so often leads me to sequester myself for months at a time. They tell me, these friends, to take it easy on myself. Be gentle. It's good advice, if hard to follow.

People used to love hearing about how I bought my horse. It was a good story, with a happy ending, which left me in the role of the cowboy. It sounded bold, and brave, and beautiful, and it was those things. But the story of selling her, though less majestic, is, I think, the braver one. Not so entertaining, and certainly less cool. But the right choice, made the hard way.

When I was with Lady, I always felt as though, together, we were more than the sum of our parts. Standing and sweating in the hundred-degree heat; riding beneath hawks' nests, watching a three-legged coyote yip at an enormous crow. On a particularly windy Saturday, while I was jangling from misprescribed steroids, we won an amateur trail obstacle competition, both of us just barely holding it together. I still have the ribbon. I still have the memory of those deer in Wyoming, of the man drifting through the air on a piece of silk puffed out like a jellyfish, waving the American flag. The blacksmiths driving hot iron until it went liquid, until it was fire. There are certain things that never leave you. Wherever you go, there they are.

A RACER WITHOUT A PEDIGREE

SARAH ENELOW-SNYDER

I was flipping through the newspaper want ads after church one day when I found Leo. He had a cribbing problem, meaning that he gnawed on the wood planks in his stall while grunting, which not only damages the stall, but also the horse's teeth and digestion, and the owner only wanted a few hundred dollars for him.

My father drove me in our small black pickup to go meet Leo, hauling an empty horse trailer behind us. I saw right away that Leo was docile and a lovely reddish-brown sorrel. I let him sniff my hand, nuzzle my curly Afro, and look me over. In my imagination he was perceptive and liked me exactly as I was, an eleven-year-old biracial girl, a string bean on my way to something other kids would later call a "Black-girl ass." I mounted up and rode him in the owner's round pen for a few minutes while my father spoke to the owner about price. When I dismounted I said that I liked Leo's style and wanted to bring him home. My father gave a warm, approving nod.

When we pulled up to our weathered gray barn, the one horse we already had came running over to see what all the commotion was about. Buck was chronically anxious, so he was pacing, high stepping, and whinnying while we unloaded

Leo from the trailer. We supervised their first meeting with halter and lead rope, and once they were suitably acquainted, we let them loose. It was originally my father's idea to have horses, and it quickly became our father-daughter activity. My father showed me how fun and freeing it was to ride, how majestic these animals were, and how human they could be. He used to say that Buck was his best friend. The way Buck came running when my father was at the gate—they seemed almost telepathically connected, and I wanted that kind of rapport with Leo. I carefully groomed and fed him, built a friendship with him, and felt the breeze on my face as we rode all over the hills near our house.

My parents had lived all over the country, studying at different universities and pursuing various jobs, and my older brother was born during a one-year stint in Texas. Then they moved up to New York State, where I was born, and we settled back down in central Texas in 1989 when I was six years old. My father was a professor of mathematical applications in political science, and he'd gotten a job at the University of Texas at Austin. Our ranch-style house out in nearby Spicewood was the same reddish color as Leo's coat and sat on ten acres. My stay-at-home mother kept a garden next to the house with tomatoes, peppers, and cilantro underneath a wooden sunshade, and on the opposite side of the house grew a sturdy peach tree.

A moonscape of white limestone surrounded our barn, with patches of prickly pear cacti and stubby cedar trees. Much of Texas, including my area, is not actual desert, though, full of grassy ranch land. In fact, many old Westerns that take place in Texas were filmed in the more dramatic-looking desert landscapes of New Mexico or Arizona. We were in a frequent state of drought—our well went dry one summer and people accidentally set land on fire by flicking a cigarette onto the ground or burning trash at the wrong time. We had outdoor cats that lived and died at the permission of wild dogs, and a parade of goats that tore up the grass by its roots so it never grew back again.

Being out in the country seemed to define everything about us. Spicewood had just a few small houses, a few trailers, and a few people in it. The fact that our house wasn't on wheels put us at the top of the socioeconomic ladder. Our area had no grocery store, no school, no library, no courthouse, no movie theater, no downtown square, nothing that would bind a community. We were forever "going into town" to do things, "town" being either Austin, which was thirty-five miles away, or Marble Falls, which was fifteen miles away. Spicewood was really more of a rural stretch between places you'd actually need to go. Austin proper was too expensive for us, and besides, my father had long wanted to live out in the country. He was raised in Los Angeles and learned to ride a horse on summer trips up in the mountains. As an adult he lived in cities of various sizes and was eventually ready to have his own little piece of land. On top of that, he grew up watching those very Westerns set in Texas. His childhood coincided with the golden age of those films, which planted the seed for his dream of having horses one day.

Our mailbox sat on the nearest four-lane road about a mile away from our house, and we had a route number, not a street number like my suburban schoolmates. Sometimes I rode Leo to pick up the mail. Until I got to college, I had never ordered a pizza because no one delivered this far into the country. We had to take our trash into town in the back of our pickup. The school bus couldn't make it up our rocky, unpaved hill, so my brother and I caught it half a mile away.

Leo and I practiced barrel racing at home on a flat stretch of dirt, loping through a cloverleaf pattern over and over as I tried to push him faster without nicking any of the three barrels. The idea was to circle each barrel as tightly as possible, which was faster than taking a wide loop around them, but the tighter I turned, the more likely it was that my leg would hit metal. Touching a barrel during a competition added five seconds to your time. Knocking a barrel over, which Leo and I were known to do, was functionally a loss. We practiced keyhole racing, which also had a very tight pattern: gallop in a straight line, then when you reach the top of the "keyhole," grind to a stop, spin around 180 degrees on the horse's haunches, and gallop back to the beginning.

One afternoon I saddled up Buck instead and tried to practice barrel racing him, thinking his speed might give me a competitive edge. I mounted up and Buck immediately started with his jitters. I positioned him at the beginning of the pattern, relaxed the reins, and ever-so-lightly pressed my legs against his sides. I was firm in my seat when he took off, but he circled the first barrel so sharply he stumbled and almost pitched me onto the ground. The second barrel I knocked hard with my leg, toppling it over. The third barrel he tried to skip entirely because he figured the point was to get back to the beginning. Buck may have been fast, but Leo felt more like my friend and teammate, like we were meant to ride together.

I took some group riding classes with a couple of girls my age from Marble Falls, which my father watched from the sidelines. The classes were led by a hearty, muscular horsewoman who introduced me to the idea of competing, and taught me the ins and outs of Western shows, the rules of barrel racing, and what the judges looked for. She also introduced me to the idea that some girls ride English, jumping and doing dressage, wearing sleek outfits that reminded me of Banana Republic ads. This happened even in Texas, where Western culture is dominant. But that wasn't for me. I wanted to be a cowgirl. Being a Texan felt like an important part of my identity, and nothing seemed more Texan than being a cowgirl: a tough, capable woman who can do anything a cowboy does. I tried to ignore the fact that the images I'd seen of cowgirls did not feature any curly Afros shoved into unaccommodating cowboy hats. But actually, once upon a time, Black cowboys were commonplace in Texas.

Texas was very late in freeing its slaves, something I didn't realize until I was an adult. There may have been a note in our textbook about it, but it certainly wasn't highlighted in class. News of the Emancipation Proclamation finally reached Galveston on June 19, 1865, more than two years after the signing, which inspired the holiday Juneteenth. I'd never heard of Juneteenth as a kid, even though it was a Texas state holiday, and over time it has gained attention nationwide. After emancipation, some former slaves in Texas found work as ranch hands, using skills they had already developed during slavery. Jim Crow gradually took effect though, beginning after Reconstruction and lasting through the civil rights movement. This rigid segregation put Black cowboys at risk for racial violence and prevented them from eating or staying in many establishments as they herded cattle and otherwise traveled for work. Over time the cattle industry evolved and Texas industrialized. As the twentieth century sped onward with its advancing technology and manufacturing, the cowboy started to feel like a thing of the past, and eventually all that remained in the popular consciousness was a John Wayne–esque white man on a horse.

Because John Wayne loomed so large in Western films, which were beloved in many parts of Texas long after their heyday, I grew up watching him. My father and I loved him in *Red River* and Lee Van Cleef in *The Good, the Bad and the Ugly*. I flipped on reruns of *Gunsmoke* after school. I had no idea back then that Wayne openly supported white supremacy, and I naively fell for his Western-frontier mystique—I couldn't imagine myself riding English.

My Western outfit from bottom to top included cowboy boots, jeans, a belt I'd made myself out of cloth and metal trinkets, a red-collared shirt, and a black cowboy hat. And my frizzy, coarse, dark-brown curls were slicked back with fistfuls of gel, stuffed into a red barrette at the nape of my neck. Barrettes were no match for my hair—I regularly bent the thin metal out of shape trying to force them closed. Part of me longed to look like an archetypal rodeo queen with long, straight, flowing blond hair and sparkling blue eyes. It seemed to me that's what a winner looked like.

Even if my physical characteristics didn't poise me for success, I assembled my outfit and joined 4-H, the nationwide youth development program. In my area, 4-H kids were mostly known for raising farm animals like rabbits, chickens, and cows, and having them judged at fairs to win prizes. But the 4-H I joined was specifically a horse club that put on Western shows.

At one of those shows, I led Leo into the ring on foot, and we lined up next to the other contestants, all 4-H kids. A judge circled each horse, checking for physical defects, looking to see if the four hooves lined up in a perfect rectangle, noting head carriage. Contestants with smooth blond hair, glittery pink eyeshadow, and confident smiles walked their horses out a few paces to show off their gait.

That day Leo was even more low-octane than usual. I repeatedly jiggled his lead rope to wake him up when the judge wasn't looking in my direction. By the time the judge came to us, I was standing tall with perfect posture, but Leo's head hung almost in the dirt. He was falling asleep, his back hoof cocked in relaxation, swatting at flies with his tail. On top of that, Leo had a not-so-muscular build and back hooves that splayed out to the sides. To no one's surprise, we didn't even place in that event. My father was in the audience, and as I walked out of the arena, he gave me notes. He said I'd worked Leo too hard before the event, so it was on me that he was so tired.

I didn't actually have a trainer, meaning someone who worked with me one-on-one to prepare my horse for competitions. I only had my weekly group riding classes, which taught basic technique. My father learned a lot from watching my classes and shows, quickly becoming well versed enough to weigh in on my performance after every event. That's the kind of father he was, always pushing his kids to perform at higher levels, no matter the discipline. He spoke very authoritatively, like the professor that he was, and I rarely questioned him.

Over time I realized that Leo and I shared a chronic problem: neither of us seemed to have the chops or the pedigree to be elite competitors, to really be taken seriously in this horse world. I couldn't do anything about our pedigree—something determined at birth that no amount of hard work can change—but occasionally I was convinced that we had the chops. After the falling-asleep incident, I took a first-place trophy in barrel racing. Admittedly I wasn't up against many other riders, but still, my father said I'd done well, which made us both smile. This called for bacon cheeseburgers at Dairy Queen.

Competition is a long-standing part of 4-H, though its origins were actually more focused on engaging young people to help modernize old farming practices. The very first 4-H club was formed in Ohio and was known as the Tomato Club or the Corn Growing Club. This was in 1902, so for decades, 4-H clubs were segregated by both race and gender. Girls' clubs in Texas offered lessons on canning, food preparation, home economics, and even sewing, while boys concentrated more on the organization's core agricultural mission. When it was founded, Black kids were not barred from the organization altogether, but they held separate meetings and couldn't compete in state or county fairs alongside their white counterparts. The Civil Rights Act of 1964 finally outlawed segregated clubs, and by the time I joined 4-H, it had been integrated for only thirty years.

But even now, change comes slowly and unsteadily. The Iowa chapter of 4-H was accused of racism numerous times in recent years, culminating in a 2019 lawsuit, according to an investigation by the *Des Moines Register*. 4-H's first Latino leader at the state level in Iowa, John-Paul Chaisson-Cárdenas, alleged harassment and systemic discriminatory practices amid his push for greater diversity.

Of course I didn't know any of this when I went to my first 4-H meeting at age twelve. I didn't know about the organization's rocky path to integration, nor did it occur to me that I would have been relegated to a different, segregated club a few decades earlier. One aspect of 4-H that has always been firmly in place is the organization's emphasis on achievement. Participants are expected to deliver results. Way back in 1919, contests and prizes became an integral part of Texas 4-H, highlighted by its girls' club motto, "To Make the Best Better."

Almost eighty years later, I entered the ring wanting to achieve a new personal best, and often walked out feeling like I had failed. But the ring gave me something that I didn't have elsewhere: a sense of knowing how I was being judged. In middle school, especially when it came to my social life, it felt like the ground was constantly shifting. One day in the courtyard between classes, my group of white girlfriends upped their usual teasing about my appearance. The leader of our group, someone I considered a friend, routinely called me names—dirty, scary, rat's nest, Medusa, Don King, Sideshow Bob—but that day she reached out, grabbed a fistful of my hair, and yanked as hard as she could. I yelped, surprised by the jolt of pain. I stared at her, holding my head, trying to understand the violence that had just been committed against me. Then she did it again, harder and harder, until I thought she was going to rip hair from my scalp. The other girls in our group cackled with the excitement of a throwdown. Soon I was grappling at her hair, but her smooth blond locks slipped right through my fingers. The bell rang and we shuffled off to class, me behind everyone else, my supposed friend announcing she had won.

The white boys in our class used to say I had pubic hair all over my head. They said I was an idiot who'd stuck her finger in a light socket. Unsurprisingly, I had never kissed a boy, and wouldn't for years.

My teachers were more diplomatic, but they could see I was unpopular. Bless her little heart, they seemed to say, with their soft eyes and weak smiles.

My hair became a focus for me. I tried to tame it into something that would slip effortlessly into a barrette, that could be held up with thin little bobby pins, that would cascade down my back, that would reflect sunlight, that looked "clean" to the girls and boys at school who told me to try using shampoo for once in my life. I took forty-five-minute showers. I studied the pages of *Seventeen* magazine like they were notes for a final exam. I tried products like Frizz Be Gone and Frizz Ease. I got my hair thinned at Supercuts. It didn't make any difference.

I was better at my actual schoolwork than the social politics that played out between classes. I had been in the gifted-and-talented program for years, although my father informed me one day that these special classes were actually kind of silly and not very rigorous. I made the honor roll and got As in my honors classes. I entered the science fair and placed fourth in my category. When I came home from the fair, my father rolled his eyes and asked why I'd done such a flimsy experiment about mares and geldings liking apples or carrots, which had no real science behind it. This had not occurred to me. I was embarrassed and hid my ribbon out of view. I was invited to take the SAT in seventh grade, but I scored so low that I didn't even come close to qualifying for the smart-kid summer camp, where my older brother went every year. I genuinely liked being in class, and thought it was fun to ride horseback, but my father's critical gaze cast a pall over these simple pleasures, and I had an increasingly hard time enjoying them.

After the hair pulling, I asked my parents to take me out of public school. I told them I didn't want to be bullied anymore. They were surprised, as I wasn't in the habit of coming home from school and telling them how exactly I'd been humiliated that day, but I knew they'd support me and wouldn't blame me for the cruelty of other children. I also knew that they prioritized education above all and would be happy to help me focus on my studies. So I spent eighth grade at home, with no classmates and my parents as my teachers.

Homeschooling was, in my case, very focused on math and English, which were the subjects my parents knew best. I woke up, got dressed, and ate a bowl of cereal when my older brother did, but instead of going off to a school campus with him, I went into my parents' bedroom to start an algebra lesson. I sat in a chair beside my father's desk and he read the lesson aloud, verbatim from the textbook, then sent me to my room to complete a few dozen exercises. When I came back, he graded them quickly in front of me. He never needed an answer key. His standards were higher than I'd ever seen from a teacher; when my older brother exhausted his high school's math curriculum by taking AP Calculus as a sophomore, my father had him take calculus again at the university because, evidently, the high school did not do the topic justice.

When I got more than a couple of answers wrong, sometimes my father said I wasn't really trying, or that I was being lazy. When I got all the answers right, we moved on to the next lesson. Math class lasted about three hours every morning. It was reminiscent of him evaluating my performance at Western shows, except that math was truly his specialty. At home he got to be both teacher and judge.

The first time I encountered a math lesson of his that was too difficult for me to understand, I was sitting alone at my little desk, dripping fat tears onto my loose-leaf paper. I knew my father would be disappointed by my inability to grasp the exercises, but even worse, he would think less of me for crying. He had lectured me many times about how we should never feel things like anger, exasperation, jealousy, or self-pity, which he said were unproductive. I dried my tears and left my barely complete, tearstained paper on his desk with a note apologizing for crying. He read my apology while I was elsewhere doing another lesson. He wrote below it that he loved me and thanked me for my honesty. I felt a little better upon reading those words, but I also understood that failing and crying were unacceptable in his eyes, and I was not supposed to do those things again.

After math class each day, I ate a peanut butter sandwich with tortilla chips for lunch and then went down to the barn to practice racing with Leo. In the afternoon I spent another few hours on English with my mother. She assigned me novels and short stories to read, and then gave me questions to answer in essay format. I was not a fast reader, but my mother was patient. Once I read an excerpt from *The Red Badge of Courage* and got a reading comprehension question wrong, but even worse, I didn't understand why the correct answer was correct. Her frustration was momentary and then I could see she was back to loving me, or maybe she'd never stopped. I always felt a little more relaxed in class with my mother, like there was less at stake. I could stumble and grow in fits and starts instead of needing to perform at my peak at all times.

My mother wasn't a horsewoman, and 4-H was not in her wheelhouse—in fact she was a bit afraid of horses—but she was a doting cheerleader, whether I brought home a ribbon or not. She seemed to understand that life was filled with ups and downs, and couldn't possibly be one long upward trajectory. My mother was a child of the Great Migration, having been born in the Mississippi Delta and then raised in Detroit, and knew the push and pull of hardship and perseverance.

In contrast, my father had all the educational, financial, and social advantages of growing up white when segregation was still a legal practice. My parents met at the University of Michigan, both lovers of books and art, but they came from very different worlds that came out in how they approached parenting.

My father didn't understand why I hung my head when Leo and I left the ring without a ribbon. He didn't understand why I got frustrated with schoolwork, or why I spent an hour doing my hair, only to emerge from the bathroom in tears. When I sulked, he snapped, "Pull yourself together!" or, "You're losing your cool!" I felt my permanent record being tarnished.

Still, I kept having victories, tiny marks of improvement. My father watched me compete in a "trail" event at a Western show that was much larger than we'd anticipated. Trail was essentially an obstacle course, and if Leo had one useful quality here, it was his unflappable calm. He would walk over bumpy bridges and go backward through mazes like it was no big deal, things that would spook many other animals. I came out of that event with a shiny, fire-engine-red, second-place ribbon, and my father ran over to me beaming. He congratulated me on performing so well against so many older competitors with their expensive animals, some with professional trainers even, and I had the ribbon to prove it. My cheeks hurt from smiling and I could feel the blood rushing through my veins. I felt a great sense of accomplishment for placing, but more than that, I'd made my father proud.

One sweltering afternoon at another large show, I was

back at the bottom of the pack. Leo and I had performed slop-
pily all day, my hair was a frizzy mess from the humidity, and
I was wiping away tears.

My father pointed to a young Mexican girl nearby and
asked, "Why can't you be more like her?" I didn't know this
girl, but I'd been noticing her all day. She had thick, dark hair
stuffed under a cowboy hat. Her distracted horse stepped out
of line, tossed his head high, whinnied loudly, and otherwise
threw every event. She placed dead last in everything. The
girl couldn't stop smiling.

She was like me, but happy.

A few minutes later she walked past me and said, maybe
to no one in particular, "Isn't this so much fun?" before disap-
pearing into the crowd.

I was stunned by her maturity, that she could survive,
even enjoy, this much failure. Or maybe she didn't consider it
failure. I was taken with the idea that I might make a mistake
without it going onto my permanent record. I wanted to let go
of all the tension I was storing inside my shoulders and finally
do something that young girls were supposed to do: have fun.

As we were packing up to go home, my father asked me
if I had what it took to continue doing these competitions. I'd
had one too many sulking moods at this point and he was fed
up. "Maybe you should just quit," he said. This was a test;
quitting was a cardinal sin in our house. I slumped down into
the passenger seat of our black pickup, exhausted from riding,
from crying, from aching to be better.

The competitions slowly faded out. After a year of homeschooling I started at a private high school and my priorities

shifted. My father seemed nervous that I wouldn't get into a "decent college," so anything that didn't resonate on applications took a back seat. I was accepted into the National Honor Society and got an after-school job as a tutor. I doubled down on my other hobby, violin, which I had also been doing competitively since I was eight years old. I ramped up my juries, recitals, and orchestra performances, because my father said that classical music made me a more attractive candidate than barrel racing. At my father's insistence I applied to a college in the northeast where several of his relatives had gone, and I got in, after which I completed my high school thesis and graduated with honors. By the end of high school, I had taken fourteen standardized tests to get into college: two PSATs, four SATs, four SAT II Subject Tests, and four AP Exams. But because a legacy weighs so much in the college application process, I'll never know if I was really smart enough to go.

At some point while I was away at college, my father sold our two horses, and then my parents divorced. I was surprised. Although there had been problems in their marriage, I didn't think they'd officially call it quits. She moved to Chicago, where her heart had always been, and my father moved into Austin. Our life out in Spicewood was over.

As a young adult, especially after college when I moved to New York City, most of my friends didn't know I used to ride. Sometimes I would casually mention barrel racing without realizing this was new information to them, and they would squint at me in disbelief over raspberry martinis in some Manhattan bar that was a world away from dusty Texas. Eventually something called Instagram came along, and with it so many digital tribes, people connecting through their common interests. One day, because I'd been following some accounts connected to prominent Black travelers, I stumbled across photos from Outdoor Afro, a group of Black outdoor enthusiasts that included horseback riders. Besides photographs of myself, I couldn't remember seeing images of a Black person on a horse before. In one of the Outdoor Afro posts, a young girl in Tennessee held on to the saddle horn with both hands while a grown-up led her horse. The girl's broad smile seemed to suggest she was experiencing something brand-new, in awe of this animal that was so much bigger than her.

Seeing that image prompted me to dig around, by which I discovered Cowgirls of Color, a group of Black women who compete in rodeos and aim to inspire young Black girls to do the same. In interviews I watched online, these women spoke openly about being outside the mainstream rodeo culture, some of them starting to ride in their thirties and forties, struggling to gain access to good trainers, hearing people say they don't have what it takes. I wondered what my childhood would have been like if I'd had access to the robust internet of today, through which I could have seen girls on horses who didn't look or perform like archetypal rodeo queens. They looked at ease with their natural hair underneath their cowboy hats, gorgeous and free.

NO REGRETS
JANE SMILEY

If I looked through my family tree, I would search in vain for anyone who was obsessed with horses, so I am assuming that it was our TV that did it to me. Black-and-white, there they were, my friend, Flicka, some guy named Roy who rode Trigger, and Fury (which I misunderstood as "Furry"). I paid no attention to the people, except to note that they got to ride and I didn't. There was one family story about my grandparents' life on a ranch they owned in Idaho. The ranch was a ways out of town, and it was said that my grandfather and his older brother had seen a storm coming, ridden out to check on the cattle. On their way home it got so cold that they got off their horses and lay down together in a mound of snow in order to not freeze to death. I had no idea if this was true, and I was grateful that my grandfather had survived, but no one ever told me the names of the horses, which was the thing I really wanted to know.

And then, around the same time that we got the TV, someone set up a pony-ride ring not far from our apartment. We stopped there as often as I could talk my mom into it. I was strapped into a Western saddle, sent at a trot around a small maze. I was hooked. Now that I could "ride," all I needed for perfect happiness was a horse friend, one who lived in our backyard or, perhaps, in the garage. I thought that this horse friend would be much easier to understand than the kids and the teachers at school—I was an only child, and for only children, social interactions can be head-scratchers. As I got to be a more sophisticated reader, I saw that I was right—horses in books were always there for you, even if they needed a little time to get acquainted. My favorite was *Silver Birch*, about a girl on a farm in Wisconsin who sees a horse running loose in the woods and uses the carrot rather than the stick to attract it, claim it, and ride it off into the sunset.

I am sure that my mother thought I was crazy, but she was patient, and she would take me to local farms that offered rides once in a while. She also sent me to a day camp where we rode in old cavalry saddles, mostly at a walk. My fifth-grade teacher ran another camp, down in the Ozarks, where my favorite trail ran across the Current River and up into the hills.

Just when I was beginning sixth grade, my mother remarried, and her new husband not only had money, he also had a kind heart, and I was allowed to take lessons every Friday afternoon at a stable that specialized in Saddlebreds, not far from our house. I was also sent to summer camp up in Wisconsin—more horses, more trail rides, a wilder landscape.

The first girl I met who shared my obsession was Dinah Stix. One of the privileges of my new school in sixth grade was that a pony lived there, and two sixth-grade girls were assigned the task of taking care of it. Dinah knew what she was doing, and sometimes I would get to be her assistant. She was more a guide than a friend, and seemed dedicated to her hobbies more than her social life. One Sunday morning in the spring, we went over to feed the pony, and there was a foal, standing by its mom, perfectly healthy. Dinah said that no one had known the mare was in foal when she came to the school, and no one had wondered why her belly was bulging, either. But we got to pet the foal and take care of the mare. Dinah told me all about real riding—Three-Day Eventing and Pony Club. I switched stables and began taking jumping lessons. But I was still an observer—of the horses, of the trainers, of the other girls who were much better in the saddle than I was, of the racehorses from across the river who wintered over in our barn (and were subjected to cruel treatments—blistering, pin-firing, and nerving to make them appear sound enough to run in the summer). In the late winter, when it was too cold or snowy outside to ride in the arena, the grooms brought the recovering racehorses out of their stalls and threw us on them. Our job was to ride them in a line up and down the aisle at a walk, which got me used to Thoroughbreds and wasn't dangerous. I continued to read horse books, and the girls and boys in the books continued to make horse friends (my favorite authors and illustrators were Marguerite Henry, Wesley Dennis, and C. W. Anderson). I was not making any horse friends, but now I could post, canter, and jump a few small jumps.

And then my parents joined a recently founded club across the Missouri River that specialized in English sports—skeet and trapshooting, horse riding. I had no idea why they joined, but I enjoyed the horses and the landscape beyond all things. That summer, a horse appeared at the barn who immediately drew my gaze—a tall dark bay Thoroughbred mare, sturdy, graceful, and beautiful. Her name was Rivertown Gal, and for some reason that I didn't understand, her owner never rode her. Dinah would come with me to the club and ride, and I was allowed to ride the bay mare. Sometimes we rode in the small arena and other times we rode all over the grounds—to the bluffs overlooking the river, into the valleys, along the edges of the fields. I was thrilled and happy, and the following spring, I proved my love of the horse after I broke my arm high-jumping in my track-and-field PE class by visiting her at her owners' house, down the hill from the stables where we rode the racehorses. I took her for hand walks around the property and the neighborhood. I was devoted to her, and my parents bought her for me. All summer, I went to Strathalbyn, the club across the Missouri, as many times as I could get my grandfather to drive me, and rode Rivertown Gal, sometimes with someone else, sometimes alone. My best memory is of a clear summer day, a trail ride when my mare and I walked up a hill out of the woods, then wandered around among the blackberry bushes, me picking and eating ripe blackberries, her snatching bites of rich grass.

I enjoyed riding her, loved her looks, and came to ride her more skillfully, but I had no sense that she was my friend. When I showed up at the barn, she would sometimes look at me, but usually she would continue to eat her hay. If I sat in her stall, she would ignore me. It was not that she had no relationship with me, it was that none of my horse books had taught me to know what that relationship, realistically, might be, and no one around me, instructors, grooms, friends, ever talked about making a connection. Human/horse interactions were practical—horses were to serve and obey and we were to get them to do what we wanted them to do, often with the whip and the spur, sometimes with the carrot. Most of my instructors were men, some of whom had been trained in the US Cavalry. My guess is that they were not encouraged to connect with the horses they rode, but to see them as instruments of war (as a point of information, the cavalry as a horse-riding battalion was disbanded after World War II) rather than friends.

I became adept enough to go fox hunting sometimes with the local hunt—Bridlespur. The day was Thanksgiving, the ground was wet. We followed the other riders over a fence, then through a field. We made a slight turn and Rivertown Gal slipped, fell, and stood up, hobbling. The fall had broken her stifle joint. A few days later she was put down. I kept riding, but in some sense, my dream had died with Rivertown Gal, and I stopped riding completely when I went off to college, Vassar, which, unlike a few other colleges, had no riding program. I thought I was done with that, and to be done with it was rather common—horse girls didn't often turn into horsewomen if they didn't have many skills, and I didn't. I wasn't going to make it as a competitive rider, and I wasn't going to make it as an instructor, and I was too tall to make it as a jockey (as if . . .). I didn't even read horse books anymore. I was a grown-up.

I didn't ride for twenty-five years. By the time I was forty-three, I had a career, three children, two houses, a love of cooking (though not of cleaning), a love of travel, a love of writing (eight published books, a Pulitzer, plenty of support) and of reading. My life was full. In the summers, my husband and I took the family to a house on a lake in northern Wisconsin where he fished and I swam or hiked about. My son had been born in the fall, two days after my own birthday. To lose the weight, I had bought myself a treadmill. It didn't work. One day in late July, I was driving around the north woods, looking for a woman who sold baby toys. I had my son in his car seat. I turned down the wrong road and there was a stable, with an arena, jumps, and turnouts. I was dumbfounded, since this was lake country, not horse country. I got my son out of the car seat and went into the barn, where I found no one, then around the barn to the arena, where a man was giving a lesson. Almost immediately, a woman showed up. They were friendly, professional, younger than I was. Since the treadmill hadn't worked and I knew I needed the exercise, I asked if I might take lessons, and they were happy to give them to me. When I showed up the next day in some outfit that I had put together, they lent me a helmet, and once I got into the saddle and began walking and trotting around, I could feel my body remembering what it was supposed to do—heels down, chin up, straight posture, light hand, sit deep. I was amazed that the skills seemed to be coming back. The next day, they put me on a fourteen-year-old ex-racehorse, dirty white, tall, good-looking. His gaits were smooth and I liked him. When I returned the day after that and approached his stall, he looked right at me and nickered. Completely unexpectedly, I had found my horse friend. I fell in love, bought him, named him Tick Tock. The excess weight that the treadmill could do nothing for came off in two weeks.

Perhaps as an acknowledgment of my new horse's wide-ranging experiences, I called him "Mr. T." I knew he had ended his racing career in Chicago. There was evidence of a suspensory injury on his left front leg. I knew he had been re-trained and sold to George and Tina, the owners of the stable. A few months after I bought him, I looked under his upper lip and wrote down the number and letters that were tattooed there and sent them off to the Jockey Club. The report they sent back was impressive—fifty-two starts, $165,000 in win-nings. What was more impressive was, according to the vet, that the conformation of his legs was perfect, and apart from the evidence of the suspensory injury, he had no soundness issues, and more important than that, he was born in Ger-many, had been trained and raced in France (I was amused that one of the races he won was called the "Prix du Nabob"), and then he raced in the US, on the East Coast and the West Coast, before ending up in Chicago. A world traveler! My old fascination with racing, and with Thoroughbreds, came rush-ing back, enhanced by my new horse's evident sophistication. Mr. T was my model—good temperament, excellent confor-mation, a stayer rather than a sprinter, followed by a pleasant retirement as a versatile riding horse. I thought I might try breeding a racehorse or two—had no idea what I was doing, and if I had, if I could go back and do it over, I would do it for one reason only—what I observed and learned about horses and their connections to humans.

The first horse I bred was born in 1996, and the last one in 2007. I chose the first broodmare I bought, Biosymmetree, be-cause she was the daughter of Big Spruce, who was known for kindness, good looks, longevity, and winning plenty of money. I went to the California farm where Biosymmetree lived; they walked her around and trotted her. She was graceful and sound. When I put her back in the stall, she turned her head, looked at me, set her chin on my shoulder. I was smitten. I bred her to another stayer (second in the Belmont Stakes, with plenty of other wins) and in the spring of 1997, the darling was born. His registered name was Sylvanshine. I called him Jackie. He was a dark bay who even as a day-old foal had a spring in his step and a curiosity about what was going on— mom was eating hay, Jackie was staring down the hill through the bars of the pen.

In the meantime, Mr. T. and I continued to connect. When I moved to California, I kept him in my backyard. He would watch me, greet me, come when I called. I would get on him and ride him idly down our road with the dogs (a Great Dane and a Jack Russell) coming along. If anything strange happened (as in, the Great Dane running into a neighbor's backyard and stealing a chicken), he would stop and stand quietly while I got off and sorted the situation out. Yes, I fell off him once and broke my ankle, but it wasn't his fault—he turned and I was out of balance and slid off. He let me ride him anywhere—into the howling cold wind down dirt roads in Iowa, into the hills and the woods in California, over jumps in arenas and on event courses. He continued to nicker when he saw me. He did not seem to like my husband, and when my husband left me and the guy I came to love showed up to fix the paddock fence, Mr. T followed him everywhere—sniffing the back of his neck, looking into his toolbox, watching him hammer the railings onto the posts. After that, he was much friendlier toward my new beau than he had been toward my ex. It made me laugh, but I also respected his perspicacity. He taught me that connection is a form of equine intelligence, a way of understanding consequences—if I was kind to him, he would be kind to me; if I attempted to understand him, he would do the same.

But let's turn that around. It seemed to me that he knew that if he was kind to me, I would be kind to him, and if he attempted to understand me, I would attempt to understand him. I am sure my childhood instructors would have sneered at this idea, but if you look at horses in herds, they evidently know that they each have jobs, that it is best to get along, and getting along involves reciprocation. At the very least, if you see a herd of horses gallop down a hill, they hardly ever bump each other—they know who's where and how to stay in their own spaces, even at a run. Bucks, pinned ears, bites, and kicks are communication devices, as are standing close to one another, mutual fly swatting, a horse curving his neck around another horse, nuzzling. A whinny can be a warning, but also a greeting.

The first thing I learned from breeding foals was that horses have nature and nurture, just like humans. Jackie was a sweet boy from the day he was born, and my favorite picture of him is from when he was about two months old. I had taken my four-year-old son with me to the barn, and he was wandering around in the outside paddock. I was in the stall. I turned to see them both looking at me through the door, and I said to my son, "Axel, give him a kiss." He reached out his hand, drew Jackie's nose toward his own face, and kissed him between the nostrils. I felt no sense of danger or dread— Jackie was friendly and calm, did not consider Axel to be a threat (Axel was a sweetheart, too). He was also too young for treats, so I wasn't bribing him—he was just being himself, the offspring of Big Spruce and Biosymmetree.

I moved Mr. T. to a nearby stable with wonderful trails, an old racetrack, decent turnouts, and then there was an episode that illustrated both his athletic skills and his attachment to me. I got to the stable, saw he was turned out in the large green space inside the training track. I went there, called him. He came running toward me. My heart started to pound, as I imagined him knocking me down, but I didn't move. He slid to a halt about two feet in front of me, his ears pricked. I petted him and gave him a carrot.

Eventually, I bred far too many horses, too many to support and enough to be fascinated by their similarities and differences. All the ones related to Big Spruce were friendly, attentive, thoughtful. One of them, Hornblower, a year younger than Jackie, spent some time at the racetrack, then was trained as a hunter. When he was about six, I sold him to a young girl, and a few months after he left, they let me know that one afternoon, she had been riding him, she had slipped to the right and might have fallen off, but he moved to the right, got himself underneath her, prevented the fall. Another time, I was leading my filly, Paras. Something in the woods spooked her and she jumped toward me, but she stopped herself, didn't even touch me. Once I was riding Jackie and he spooked. I fell off, landed on my back. When I stood up and put my hand on the spot that hurt, he came over and sniffed it. I took my hand away. He nuzzled the spot.

I had other horses, horses I'd bought, who would buck

me off and run away, but the ones I'd known from foalhood would always come to me, check on me. I also watched how they related to one another. For many years, Jackie was turned out with three of his female cousins. It was a big pasture and they enjoyed running around. He seemed to be especially attached to Essie, who was a year younger than he was. I would exercise them together, mounting one of them, sometimes bareback, then leading the other up the hill to the arena. While I rode the first one, the other one would wander around the arena. I would then dismount, climb on the other one, and ride that one while the first one wandered around. They were always calm and agreeable together.

All of the horses I bred who were related to one another were active, attentive, excellent movers, somewhat spooky, capable of learning, sensitive. I stopped using spurs or whips (I changed from a whip to a baton—a flexible stick with a red ball on the end that didn't hurt if I used it, simply made a sound). It seemed as though they wanted to please, but were occasionally overcome by their own Thoroughbred liveliness. They had their preferences—Jackie loved to jump. Paras, who I ride now, would follow Jackie over some free jumps (I could also hold her, point to my husband on the other side, say, "Get the cookie!" and let her go, and watch her jump the jump and go straight to my husband), but if my trainer or I tried riding her over a jump, she would buck like mad.

When I tried to teach them voice commands, they learned them readily—not just "Whoa," but "Halt," "Walk," Trot," "Canter," "Trot UP," "Where's Dad?" (which meant "Look for Jack," my husband). Paras learned "Heel," "Ah-Ah" (which meant stop acting up), "Wait," and to come if I called, "Per-eSTROIka!" She also learned "Go to bed," which meant "Go into your stall." Ned, the last horse I bred, born two years after Paras, learned to come if I whistled a certain set of sounds. Ned also learned "Go left!" and "Go right!" Both of them knew more commands than any of our dogs.

But it took me a long time, and a new trainer, to teach Paras to be a good riding horse. It turned out that the prob-lem wasn't hers, it was mine—I was giving her inconsistent signals, not understanding her nervousness about being out of balance, and her preference for a bitless bridle. Once my new trainer had analyzed what the problems were and I corrected myself, it was evident that she knew what she was supposed to do if I was able to give the proper signals. That didn't mean she stopped being opinionated—she still hates to load into a trailer, she still doesn't like new things, and she still won't go in the lead on a trail. But if we want to maintain a friendship, which is what I want to do, then it is my job to understand her quirks and allow them, just as I wouldn't serve roast beef for dinner if my vegetarian friend was coming.

I have won a few ribbons over the years and sold a few horses, but the primary thing I have gotten from my invest-ment is the sense that they know me and they connect with me, the very thing I wanted most when I was young. What I didn't know I would get was the opportunity to satisfy my cu-riosity about the nature of horses—their individuality, their intelligence, their readiness to communicate, the pleasure they take in some of the jobs we give them and the annoyance they feel about some of the others. As domesticated animals, they are not our servants, they are our collaborators, and they have feelings, or, let's say, opinions, about collaborating.

The other thing I've learned that the horse books didn't mention was that they all die, like Rivertown Gal, but not always as suddenly. The average life span of a horse is about a third of the average life span of a human, which means that I must understand that my dear friend will show me what death feels like and what it means in ways that I did not understand when I was fifteen. When I am riding my horse, we are moving through the world together, and one of the best forms of connection is the rhythm of my horse's movement as it ripples through my own body, or my horse's response to how my body moves—I can think of slowing down, and almost before I sit deep and touch the rein, my horse slows down. When I used to jump, what I enjoyed the most was how, just before the jump, my horse's body and my body gathered energy together, and then put it into that graceful arc. My memories of the horses I've loved and lost are not only visual, they include all of my senses—the trees I saw on the trail ride, the sense of speed as we galloped. Once I let Mr. T. go around that practice racetrack at the speed he wanted to go—to signal he was allowed to do so I bridged my reins and leaned forward. He went so fast that my eyes watered. I remember the sound of the even clip-clop of a forward trot, the smoothness of Paras's coat when I pet her on the neck, the smell of the barn. I love taking walks, and I love the natural world, but I feel lucky to have experienced much of the natural world from the back of a horse, and to be able to cherish those memories after the horse departs.

DAREDEVILS

MAGGIE SHIPSTEAD

The summer I turned five, in 1988, my parents drove me and my older brother and our two golden retrievers from California to Michigan. My mother's mother was dying of pancreatic cancer in Ann Arbor. Somewhere in the Great Plains, left alone in a parking lot while we ate lunch, the dogs destroyed the interior of our Nissan station wagon. Of the seat belts, they left only jauntily waving flaps, and worse, they gnawed away the carpet in the back seat footwell to expose a hot metal pipe that would burn an unwary child's feet. But it was the '80s, so we duct-taped what we could back together and carried on.

In midsummer, after my dad had gone back home to work, my mom proposed a weekend outing to Niagara Falls, just her and me and my brother. She must have desperately needed a break from illness and grief, and she told us it would be a fun adventure, which, for a timid child like myself, was not a persuasive sales pitch. I saw no need for fun adventures. Adventures were inherently not fun. I was safe and comfortable at my grandparents' house, thank you very much, and so even though I had no idea what Niagara Falls was and little concept of waterfalls in general, my instinct was to react as if I had been asked if I wanted to be ritually sacrificed.

We went anyway. I was so obnoxious in the car, my mom

pulled over somewhere in Ontario to swat me on the butt, but when we finally exited the highway onto a garish strip of wax museums and haunted houses and candy emporiums, my mood lifted. Niagara Falls, it turned out, was indeed fun. We visited tacky attractions and walked behind the Falls and went on a boat that nosed into the mists while everybody crowded on deck in blue raincoats to commune with the roar of the Great Lakes plunging off a cliff.

The IMAX theater had an exhibit of the barrels and giant rubber balls and glorified tin cans that daredevils (a new word for me) had stuffed themselves into before going over the Falls. Surprisingly, most survived. The first ever was a woman named Annie Edson Taylor. She went over in 1901, on her sixty-third birthday, though she claimed to be twenty years younger. She'd hoped the stunt would make her rich and famous. It didn't. One of the most recent was a guy named Karel Soucek in 1984, who'd taken the plunge in a red cylinder painted with his name and the not particularly catchy motto, "It's not whether you fail or triumph, it's that you keep your word . . . and at least try!" He'd survived the Falls but was killed seven months later when the same cylinder was dropped from the ceiling of the Houston Astrodome with him inside and missed the tank of water it was supposed to land in.

I peered into the battered homespun containers and recoiled at the same time I felt a submerged pull of fascination. Why would people do this? I tried to imagine being shut inside a barrel or rubber ball, feeling the pushing, bobbing river give way to falling. In most of the containers, the daredevils couldn't even see out. They just had to hope their plunging vehicle wouldn't break up on impact, as happened to a man named Red Hill Jr. in 1951, or that they wouldn't be held under the down-rushing water and suffocate, as happened to George Stathakis in 1930, or that their ballast wouldn't break through the bottom of their barrel and take them with it, leaving behind only an arm, as happened to Charles Stephens in 1920. Breathless placards and canned narration played up the idea of the Falls exerting an irresistible magnetism. Risk itself was the allure, as well as the hoped-for exhilaration of survival.

There was something titillating about the helplessness of the people going over the Falls, something intriguing about the idea that an act requiring no skill or knowledge, only pure, foolish bravery, might somehow change the people who survived it, let them glimpse something at the dark, fluttering edge of mortality that could not be encountered any other way. Freud theorized that we all possess instincts toward life and creation, which he designated as Eros, but also toward death and destruction, or Thanatos, which underpins the popular concept of a death wish. Modern researchers, however, like psychologist Kenneth Carter in his book *Buzz! Inside the Minds of Thrill-Seekers, Daredevils, and Adrenaline Junkies*, have found that people who are willing to take risks in pursuit of powerful experiences and sensations have no more of a desire to die than those who prefer more mild thrills, like knitting. Rather, risk- and sensation-seekers seem wired to seek extremes in order to feel alive. As a child, I didn't know anything about any of that, but in considering the daredevils, for the first time I found myself curious about what might lie on the other side of risk.

But going to Niagara Falls did not make me bold. Horses did. Or, bold-ish. Within reason.

I don't know when or why I decided I wanted to ride horses. The impulse feels innate, preverbal, somehow pre-knowing-horses-exist, as essential a part of myself as my contradictory lack of gutsiness. My mother had wanted to ride as a child, but her parents couldn't afford lessons. Maybe I inherited the trait from her, some obscure allele expressed as an urge to cultivate herds of plastic Breyer horses and check out every library book with a horse on the cover. When I was seven, we moved to a different town, and I was bribed to go quietly with the promise that I could take riding lessons at a barn near our new house. So began the long, slow morphing of small me, beaming in rubber boots aboard a plodding school horse, into teenage and early-twenties me, stern in professional action shots taken midair at horse shows as I jumped my increasingly fancy mounts over fences that were three feet, then three and a half, then four.

Fear was a part of riding from the beginning. I was afraid when the horse scooted or bucked or spooked, afraid when the jumps got higher, afraid when my trainer suggested I go for a trail ride because you never knew what was going to happen out there among barking dogs and lawn sprinklers and bushes that abruptly disgorged speeding rabbits or flapping clouds of doves. None of these things posed any physical threat to a horse, but horses are bad at assessing danger. Some people think it's a sign of stupidity that a thousand-pound animal might leap into the air and race away when, say, a breeze rolls an empty paper cup along the ground, but, in terms of evolution, the reaction makes sense. As a prey species, prehistoric horses' survival hinged on reacting swiftly to unexpected movements and sounds. Horses who ignored the rustling grass as the sabre-toothed tiger crept closer ("Chill out, guys, I'm sure it's fine") were less likely to stay in the gene pool than those that bolted. This was an animal I could relate to.

I became a proficient, experienced rider, but I wasn't particularly talented. I never developed the unflappable confidence the best equestrians have, nor did I burn with ambition to jump ever higher. At any level of riding, you can be badly injured or die from a fluke accident, but as the fence rails creep up, you must be ever more precise and decisive in order to get around a course safely. If you allow your body to transmit uncertainty to your horse, he might refuse or run out, launching you off his back and onto the jump. If you get picky and nervous, you might arrive at the base of the jump too soon or too late, forcing the horse to take off from an awkward distance, and the bigger the jump, the more likely you both are to find yourselves tangled in a collapsing heap of heavy, clattering poles.

Getting better, though, meant pushing myself, and gradually I found my way to an ethos of Doing It Anyway. When my trainer raised the jumps, I'd tamp down my dread and adrenaline and pick up the canter and go. Usually things turned out fine. Being afraid, I slowly came to understand, doesn't necessarily mean imminent doom. It's just a feeling, an evolutionary warning, and with practice you can learn to set it aside. I'm not sure why I felt compelled to push myself quite so hard. Why didn't I just top out at a fence height where I was comfortable and keep trying to improve my technique? After all, what I loved about riding wasn't the endless boundary pushing but rather the collaboration with the horse and the tantalizing hope of a perfect ride. Pride drove me, I suppose, and my competitive nature, my inability to reject a futile quest to keep up with the equestrian Joneses. I suspect that, for many or most elite riders, jumping higher doesn't mean Doing It Anyway but rather Doing It *Because.* Sensation-seeking, as a personality trait, is complicated to assess, but to excel in a dangerous sport, it's helpful to have the kind of brain chemistry that makes physical risk its own reward, as opposed to an inner obstacle that must be surmounted along with the actual jump.

I stopped riding thirteen years ago, when I was twenty-four, well after it had become clear that not only would I not be able to fund a wildly expensive sport in the foreseeable future but also that I wasn't on the verge of the kind of settled life that allows for the care of a large animal and a commitment to steady training. And—the question lurked—what would that training be *for*, anyway? I loved horses, yes, and I had ridden for a long time, but at the end of the day, I wasn't all that good, nor did I know how to ride just for the pleasure of it. After I aged out of the cutthroat world of junior equitation, I found myself competing in amateur divisions mostly against middle-aged women who had ridden as juniors and then returned to the sport following long hiatuses spent establishing families and/or careers. My impression was that most of them were somehow able to devote staggering quantities of time and money to riding while also staying pretty mellow about the whole enterprise. They seemed to be having fun, whereas I didn't know how. Riding wasn't just a hobby to me, but I didn't know what it was. Then, all of a sudden, it didn't matter. I quit, and horses became my past.

There's a cultural narrative floating around that goes beyond acknowledging the special fascination horses hold for (some) girls and instead dismisses a love of horses as inherently girlish—the fleeting and cutely bemusing preoccupation of a harmless demographic. As girlhood is temporary, so, too, are horses something to be grown out of: stand-in love objects for the boys and men that will eventually supplant them, outlets for nascent romantic urges that will inevitably redirect toward wife- and motherhood. This is, of course, a patronizing and heteronormative take and also one that ignores the boys and men who love horses, but its assumptions are more ingrained and pervasive than they have any right to be. "Better horses than boys," parents say about their rider daughters, as though the two are mutually exclusive, as though the girls at my barn weren't always enticing the few boys in our midst up to the hayloft to play truth or dare. When a love of horses gets conflated with romantic love or sexual passion, the underlying implication is that girls aren't capable of really knowing what they're interested in or what they want to do with their bodies. The minds of girls are treated with condescension, as muddles of misdirected impulses and bewildered horniness, gauzy pink zones of confusion where the desire to ride a horse is indistinguishable from wanting to have sex with a man.

I blame Freud. Well, to be fair, I blame the way the idea of the subconscious is often turned against people, used to sexualize and shame and to undermine individual agency. Search the complete works of Freud and you get 385 hits for the word *horse*. Most of these are in reference to "Little Hans," one of Freud's most famous case studies. Hans, a five-year-old, had seen a horse fall down and die in the street while pulling a heavily loaded furniture van and afterward developed a phobia of horses, particularly those pulling heavy loads. Freud, through a long correspondence with the boy's father, concluded that Hans associated horses with his father because both had large penises, or, as Freud puts it, "widdlers." (The word *widdler* appears 115 terrible times in Freud's works.) Via an elaborate chain of frankly silly inferences, Freud determined that the root of Hans's phobia was a fear his father would castrate him in punishment for his sexual impulses toward his mother. In Freud's view, exhaustingly, sexual anxieties and urges saturate every person's entire emotional life.

Anna Freud, Sigmund's daughter, took things one step further when she wrote that "A little girl's horse-craze betrays either her primitive autoerotic desires (if her enjoyment is confined to the rhythmic movement on the horse); or her identification with the caretaking mother (if she enjoys above all looking after the horse, grooming it, etc.); or her penis envy (if she identifies with the big powerful animal and treats it as an addition to her body); or her phallic sublimations (if it is her ambition to master the horse, to perform on it, etc.)." I can only speak for myself, but I've never been sexually stimulated by riding; I love tending to animals but have no desire to be a mother; I'm confident I don't want a penis, even though I deeply resent the disproportionate power men wield in the world; when I ride, I don't want to dominate horses but to collaborate with them. I like connecting with a living thing, being close to another body without that closeness carrying the significance it does with a human. Why can't the love of horses be about *horses*? After all, sometimes a horse is just a horse.

To me, a more likely explanation for Little Hans's unease, rather than anything to do with widdlers, would be the trauma of witnessing, at close range, an animal die suddenly and in distress as a result of forced service to humans. Part of what makes horses compelling is the tension between their size and strength and their fragility (those slender legs!). Little Hans, long before he was ready, experienced a confrontation with a large animal's mortality that probably served as a terrifying revelation about his own, as well as perhaps his own capacity for cruelty. (When Hans's father asks him if he would like to whip and beat horses, the child replies yes and tells a made-up story about beating a horse. Hans sounds awful, honestly, but maybe his relish for sadism is one way of trying to release himself from the burden and discomfort of feeling compassion for the dead horse.) I suspect that children, and particularly girls, are drawn to horses as much because they recognize and relate to the animals' vulnerability as by fascination with horses' power. The union of horse and rider is a dynamic bargain: in exchange for borrowing the horse's speed and power, the rider assumes the risk of losing control, of falling, of accidental death, of being responsible for injury to the horse.

Though I held out longer than most, when I stopped riding I was still, at least on the surface, enacting the same old clichéd, told-you-so story: girl loves horses, girl grows up, girl leaves horses behind. I was living in Iowa City, in my second year of an MFA program, and my parents were paying to keep my horse at a barn forty minutes away. I was only riding once or twice a week, which was not enough, and I was getting embarrassed about the ostentatious privilege of owning an actual horse when most of my classmates were living independently off their modest fellowships and stipends. I hadn't stopped loving horses or lost interest in them. There was no boyfriend jealous of my time and affection, no white-picket-fence dream eclipsing my dream of a perfect round in the hunter ring. The bottom line was that to keep riding would have been impractical and unaffordable, and, also, I was entering a stage of life where I wanted to be mobile and unencumbered. Most high school and college athletes take huge steps away from their sports after they get spat out into adulthood. Riders aren't so different. Time and resources run short. Skills and fitness fade. Stakes lower, and questions arise about the point of continuing.

When I finished my MFA, I had no idea what I was going to do long term, careerwise. I knew enough not to bank on supporting myself with my writing, but I got a small fellowship for the year after I graduated: enough money to go somewhere for a while and make a start on a novel, certainly not enough to keep a horse. Even if my parents had been willing to keep paying for me to ride, I didn't want them to. I didn't want to be beholden, nor did I want to be a drain on them. My life wouldn't be my own until I paid for it myself. We sold the horse. I packed my riding stuff into my tack trunk and put it in storage.

I didn't get on a horse for eight years.

———————

When I look back at who I was then in relation to who I am now, I can feel, in a wordless, almost tactile way, the continuity of my essential, innate self (the child who cries at the idea of a new experience), but I also see large, improbable shifts in outward circumstances and trajectory. That Niagara Falls summer, I would have seemed an unlikely candidate to become a travel writer. A novelist, sure, safe inside my hidey-hole, but someone who takes trips interesting enough to write about? Hard to believe. But that's what has happened: I've established a side hustle in travel journalism, specializing in adventure-oriented trips. A photographer traveling with me on assignment said that his editor had remarked that he only ever had to buy medevac insurance for freelancers on *my* stories. I was flattered. Change often seems inevitable in retrospect, when that part of the narrative has already been lived and worked through, when an undeniable outcome trails the connected dots of causality behind it. But change is often imperceptible while it's happening.

The idea of epiphany looms large for students of fiction writing, or at least it did when I was a student. We—those in my writing workshops—usually seemed to accept that the best, most artful goal a short story could accomplish was to lead a character convincingly to an epiphany. The action of a story, ideally, would be a sort of Rube Goldberg machine that had as its payoff a realization that permanently and pervasively realigns the character's internal reality and, as more than one of my teachers put it, changes everything, forever. (For instance, a man might realize that his fear of horses is actually about widdlers. That, in fact, *everything* is about widdlers.) One problem, though, is that coming up with such a realization places a demand of profundity on writers that is so difficult to achieve, the ambition itself can become paralyzing. As a result, fiction workshops are awash in short stories where the main character, inscrutable to both reader and writer, stares off into the distance, thinking something unknowable that has, in some unarticulated way, changed *everything*.

In life, an epiphany is less often an all-encompassing end in itself than a trigger for a small change or progression and might only be apparent or emerge as important later. An epiphany might be a turning cog in a process of growth that requires effort and patience. A small epiphany (Niagara Falls is a pleasant destination; my instinct to stay in my comfort zone has the potential to rob me of meaningful experiences) may change behavior in ways that facilitate more realizations along the same theme (jumping the horse over the jump despite being nervous gives me a sense of accomplishment), which may, over time, equip one to take larger risks, and all of this might eventually begin to form a pattern, maybe a practice, that might slowly change everything.

With hindsight, I can trace a path from one risk to another and see how each facilitated the next, even though, in the living, life tends to feel more like a series of blind corners and baffling junctures: the maze as seen from above versus the maze from the lab rat's perspective. There were the round-the-world plane tickets my friend Miranda and I bought after college, the dangerous bus ride we took into the Himalayas, the extraordinarily foolish decision I made to let strange men drive me alone out of Delhi (it's a long story) that turned out okay but still haunts me with what-ifs—one risk too far. There was my choice to use my little fellowship from Iowa to spend eight months, the off-season months, on Nantucket, where I knew no one and never met anyone, while I wrote the first draft of a first book. In such deep solitude, time moved strangely, cyclically, bouncing and rolling like a tumbleweed. I got weird and anxious and overly rigid about my daily routine, but I emerged from the island as though from a crucible, newly unafraid of loneliness and boredom.

Once I understood that being alone wouldn't kill me or even make me miserable, I was liberated to go where I wanted, when I wanted. That novel I wrote in Nantucket, against all odds, paid for my life for a while. I'd ended a long-distance relationship I'd felt so oppressed and smothered by that I couldn't imagine ever wanting another. I started traveling alone. I wrote my second book quickly, in five months I spent abroad, not really talking to anyone. After that, I went to New Zealand and drove around the South Island by myself. After that, I lived on the side of a mountain in Montana with my dog for a couple months. Of course, these choices were facilitated by a densely woven safety net of privilege. During a period when I couldn't decide where to live, my parents let me hang out at their house for months at a time, and they took care of my dog when I was away. If I had failed or my income simply dried up, I knew I would have somewhere to go and time to pick myself up again.

Little by little, my life began to diverge from what I had thought it would be. The siren song of the white picket fence hadn't been what pulled me away from horses, but, when I was in my teens and twenties and contemplated the future,

I always imagined myself married, settled, categorizable in a social category that inspired less unease than "single woman." Sometime around my thirtieth birthday, though, I stopped imagining that life, not all at once but slowly, without even noticing, until one day the vision was gone. Nothing concrete came to replace it, no clearly defined alternative. Instead there were questions and fleeting wishes, flickers of possibility. I had not been able to predict my past. Why should I be able to predict my future?

After my second book came out, I was offered an assignment writing a profile of a ballet dancer for a travel magazine. When I'd finished, in what would turn out to be a consequential fit of boldness and also one of those unexpected inflection points in a life, I asked the editor if I could start pitching story ideas. My first pitch that went through was a trip to the rugged subantarctic islands south of New Zealand that I had noticed in a guidebook when I'd visited the country two years before. Getting to those islands meant spending almost two weeks on a small ship in the notoriously stormy Southern Ocean. I was nervous, but my nervousness faded once we left port. The twenty-foot (and bigger) swells I'd feared were a reality, but the reality was, somehow, not frightening. The staff and crew of the ship was accustomed to the huge seas and paid them little notice, casually inclining their bodies with the motion as they went about their business, only bothering to hold on to something when the ship rolled to thirty degrees or so. I imitated them as best I could and was lucky not to be seasick, just sleep-deprived from sliding up and down my bunk. If the sea had been flat, I realized, I would have been disappointed.

The man who owned the expedition company was on board, running the trip, and after a week or so, I began to suspect something was brewing between us even though we were so different we had difficulties signaling interest. In the hour before dinner when all the passengers hung out in the ship's bar, he would wordlessly plonk down an extra bowl of coveted and strictly rationed potato chips at my table and walk away. Mostly it was just a feeling. I was skittish and verbal; he was gruff and intense, given to staring out at the sea with such concentration he looked almost tormented. At first I couldn't decide if I was into him; then I decided I was. He was three decades older and fundamentally alien, seemingly unafraid of everything I found most daunting. He'd spent big swathes of his life camping in the bush and contending with the massive, frigid swells of the Southern Ocean and snowmobiling across Siberian sea ice. He knew about birds and plants. Harsh, inaccessible places tugged on him, and he followed his geographical yearnings not recklessly but not fearfully either, with a single-minded tenacity.

Nothing happened on that trip, but we kept in touch. Before long, via email, we'd bluntly acknowledged our attraction, plonking it down like another bowl of potato chips, and decided we'd like to see each other again. Writing to me via satellite phone from his ship, he always included the wind speed and direction. *Wind 30 knots, WSW. Thinking of you.* My friends referred to him as O.S., Old Salt.

For our first date, he invited me on a five-week-long commercial trip he was leading to the remote Ross Sea region of Antarctica. Going was a massive gamble—there was no escape route if, three days in, it turned out we didn't like each other or had bad sexual chemistry—but I went anyway. I wore a staff uniform and tended bar in heavy seas, helped get people in and out of inflatable Zodiac rafts, sat bird-watching in the bridge beside O.S. *Royal albatross. Wandering albatross. Sooty shearwater.* I saw icebergs as big as cities. I saw the abandoned huts of explorers, frozen volcanoes, hooting masses of nesting penguins. In the evenings, I went to O.S.'s cabin and got in his bunk with him while the midnight sun streamed through the porthole. In a love letter he left under my pillow, he wrote that I was "like a wildflower—a beautiful subantarctic megaherb just coming into blossom."

You won't like his body, my mother had warned. *It won't be what you're used to.* And it wasn't, but I did like it. My friends and family had not been wrong to be skeptical of my choice to go on a five-week, 5,500-nautical-mile first date with a stranger, but when my gamble paid off I felt more than lucky: I felt truly bold.

Perhaps predictably, my relationship with O.S. didn't translate well to dry land. We were together less than a year. With him, I'd felt hyperaware of the limits of my competence. I wasn't tough or rugged. I was *the girlfriend.* He was interested in my improving my skills, but mostly insofar as improved skills would help slot me into his life. Once, when we were lazing around in his bunk, he said in an offhand way that if I liked being on the ship I could earn a basic maritime certification, which would make it easier for me to be a regular staff member and travel with him more. I liked that idea, but I didn't like how, one time, when we spent a whole afternoon and evening with friends of his, I wasn't included in the conversation. It never even came up that I was a writer. Because I was so much younger than O.S., I was assumed to be an accessory, maybe even a gold digger. I expected as much from strangers (in public I often felt protective of him, tried to demonstrate with my body language that I was an active participant in our relationship), but I sensed that his friends' lack of curiosity was intended almost as a mercy, sparing me from having to explain the existence I'd wanted to escape badly enough to take up with someone with whom I had so little in common.

It's not quite that I wanted to *be* O.S.; I was realistic enough to know I wasn't going to spend my life wandering the seas and frozen wastelands, accruing esoteric expertise. It's more that I realized I couldn't emulate the parts of him I admired— his dauntlessness and his self-sufficiency—while being his girlfriend. By the time we broke up, I'd started purposefully pitching more travel stories that would give me opportunities to grow, possibly change. Slowly I was becoming a timid person outwardly leading the life of an adventurous person. I was afraid of deep water, but, for a story, I swam in the open ocean with humpback whales. I learned to scuba dive so I could go spearfishing with a famous chef. I fretted about plane crashes, but I flew in a small plane over the Arctic Ocean as the cloud ceiling pushed us lower and lower, toward the black water. *Was* I an adventurous person? Or an impostor of unusual dedication? Or neither? Or both? When does what you do become who you are? Or doesn't it?

When I got back on a horse, it was on a wind-whipped

estancia in Chilean Patagonia. The animal was short and sturdy, with a huge long mane heavy as a doormat. I was on an assignment I'd pushily campaigned for: a multiday horse trek through Torres del Paine National Park with my best friend Bailey. A female buddy story, I told my editor, about rediscovering our equestrian youth. The night before we started, Bailey and I lay awake, wondering if we still knew how to ride. When we'd met on our college equestrian team, we'd both been at our peak skills. She not only rode extremely well but managed to be a steely competitor without seeming to sacrifice her generous good nature. The year after graduation, we'd shared an apartment with two other girls and, somewhere in those aimless months, among many hatched and forgotten plans, we'd decided we would one day ride in Patagonia.

It turned out, eleven years later, we could still ride functionally enough, but this riding was like nothing I'd ever done before. The stirrups on our fat sheepskin saddles were positioned so our legs stuck out in front of us as we cantered across grassy flats and followed beret-wearing gauchos up steep, muddy mountains while wind and rain abraded our faces. The horses picked their way through fallen timber and latticed root systems. They slid down granite slopes so sheer and bare I could smell their shoes burning against the rock. They waded into cold, chest-deep water without hesitation. This was the stuff of my younger self's trail-riding nightmares, but I was starting to understand something I hadn't internalized during years of riding in circles in enclosed arenas: horses can take you out into the world. Humans have recognized this potential since prehistory, but somehow, for me, it was an epiphany. For the first time, I was experiencing riding at its most utilitarian and gloriously primal, getting from one place to another, passing through a magnificent landscape without engine noise or crowds, aboard sure-footed animals whose stoic hardiness made my heart swell with that old horse love. Every night, after hours on the trail, Bailey and I collapsed in our tent, groaning in pain, naming all the body parts that hurt—knees, ankles, butts, backs—and also exchanging fantasies about how we could keep riding after the trek. We wondered if we could find a way to take lessons, to get back into it.

Then we went home, and our lives resumed, horselessly.

I didn't ride much for more than two years after that, until I got myself assigned a story about a horseback safari in Botswana's Okavango Delta. The riding requirements were:

1. be able to ride all the gaits
2. be able to post the trot for ten minutes straight
3. be able to gallop out of danger.

Danger meant a run-in with big game, so what rule number three really meant was that you couldn't fall off. I decided to take a couple of lessons before I left just to make sure I wasn't too rusty, and promptly fell off, landing on a jump. My ankle blew up to the size of a grapefruit, and my foot turned black. It was mostly healed by the time I got to Botswana, but I hadn't been able to take any more lessons, so my first time mounting up in a wild floodplain crowded with elephants and hippos and lions and Cape buffalo was my first time mounting up at all since landing in the dirt.

I didn't give up horses for domesticity. Horses gave me the nerve to create an unorthodox life. In 2019, before COVID-19 would put a stop to things, I spent a total of five months away from home, mostly alone, and while I would say without hesitation that I love to travel, I will also admit that I feel dread before every trip. I'm so comfortable in my little bungalow in Los Angeles, with my little routine, in my city with my friends and the hiking trails I know and the coffee shops I like. Why should I leave these things? Why should I take risks when risks are scary? But I've become subject to an experiment of my own design, conditioning myself through rewards yielded by past risks. I don't mean the reward of continuing to live after going over the Falls in a barrel, nor the thrill-seeker's rush of celebratory hormones. I mean the reward of an expanded life, of possibility. Boldness is not an absolute, like a time for running a fixed distance. It means something different to all of us. I'm still afraid of most of the things I've always been afraid of, but the fear no longer obscures possibility as much as it once did. Fear is about a loss of control, but boldness is also partly about surrender. I think I am seeking surrender. I know I am seeking awe.

Yes, we startled some lions out of the bushes in Botswana, but they ran away from us. The scariest animals turned out to be elephants, those gentle, empathetic, matriarchal animals that, like horses, move me with their size. Elephants in general are protective of themselves and their young, and plenty of the individuals and families in the Okavango Delta are refugees from poaching in Zimbabwe and have even more reason to be unfriendly. Some people speculate, too, that elephants might have a collective, passed-down memory of being hunted from horseback. So I don't begrudge the enormous bull who lifted his head and flapped his ears at us, considering a charge. Our guide, in front of me at the head of the line, studied him calmly, assessing his body language. I could feel my pulse in my neck, my wrists. The animal was so big and so close and so obviously unhappy with us. What would come of this latest risk? Where would it lead?

"Turn your horses," the guide said quietly. "And get ready to run."

UNCONQUERED
BRAUDIE BLAIS-BILLIE

Growing up on the Seminole Tribe of Florida's Hollywood Reservation, everyone called us "the Frenchies." This was because, since I was around eight years old, my mother—a French-Canadian woman conspicuously named France—raised me and my two younger siblings as a single parent on the Rez. Blond-haired and blue-eyed, she stood out at every basketball game and community holiday dinner. "That's your mom?" my neighbor asked when she picked me up from an after-school playdate down the street. "I look more like my dad," I offered. It was true. My father and I had the same almond-shaped eyes and sleek brunette ponytail hanging down our backs. Reserved yet mischievous, he oscillated between cracking jokes and reading World War II books in his room; from him, I inherited my quiet, curious nature.

Most times, "the Frenchies" felt like a warm recognition. Our ekòoshes cooed the nickname with love when we walked through their doors for birthday cakes or sleepovers with the cousins. Other times, it hung in the air like something rotten. When used by certain people on the Rez, "the Frenchies" was a sharp-edged epithet that meant "not like us." It also complemented the less creative nickname we heard in Canada whenever we visited my mother's Québécois family: "les

Indiens." As I got older, each label became more uneasy for different reasons.

My siblings, Tia and Dante, and I were born without a Clan. A traditional extended family unit, Clans are matrilineal, meaning that because our mom is white, we'll never have one. It meant that "the Frenchies" were not invited to partake in ceremonial medicine like the Green Corn Dance, that we were always considered "half," that we were not wholly welcomed in the place we call home. Because our Seminole father, July, was absent from our childhoods, his mother—our empóshe, Grandma Hannah—would visit on weekends to feed us candy and teach us the culture that France couldn't and July didn't.

"We're unconquered people," Grandma Hannah said. Like my mom, she stood around five feet tall, and she was commanding despite her soft-spokenness and salt-and-pepper hair. On the days Grandma Hannah picked me up from my Head Start preschool program, she told the legend of our ancestor, Chief Osceola. Osceola was a warrior who led Seminole resistance forces against the US military in the nineteenth century. He famously drove a dagger through the 1832 Treaty of Payne's Landing, refusing to concede the Tribe's Florida lands in exchange for "Indian Territory" west of the Mississippi. It was a heroic, radical act that quite literally made it possible for us to be here today, thriving in Florida. Through her eyes, I saw our little neighborhood of brown relatives and cookie-cutter homes built by the government's Department of Housing and Urban Development as a blessing.

It was Grandma Hannah who brought me to my first rodeo. From an early age, I passively absorbed rodeo culture as an integral part of Rez culture; it was evident in the rusty horse trailers parked in neighbors' driveways and the Tribal Councilmen who donned cowboy hats and Western boots at every public event.

As an adult, years after Grandma Hannah's introduction, I learned that the connection between my people and horses comes from a long line of cattle keeping. Before moving into gaming in the 1970s—both my dad and Grandma Hannah worked stints at the Seminole Classic Casino—the Seminole Tribe of Florida provided for our people through enterprises including citrus groves, tourism, tobacco shops, and, most prominently, cattle. Over the past century, the Tribe has risen as a major player in the cattle industry. Seminole Pride, the Tribe's beef brand, is a sizable contributor to Florida's beef cattle herd—the tenth largest in the nation, according to a 2016 report by the US Department of Agriculture. When I drive through any of the more rural Seminole reservations—like the winding fourteen-mile Snake Road that cuts through the Everglades to the Big Cypress Rez—I see horses and cattle dot the swampy pastures.

As a child, I wandered the rodeo grounds sweating through the South Floridian heat in my traditional patchwork skirt. I watched in awe as Tribal members competed in events like barrel racing, bronc riding, and "stray gathering," in which a team on horseback captured runaway calves with lassos. "One day you'll be down there," Grandma Hannah declared as we sat in the bleachers chewing warm frybread and sipping cold ponche, the Mikasuki word for "soda" that I thought was universal until I started going to my white friends' houses. She noticed my ease around horses, that counterintuitive confidence required to handle a potentially lethal animal. My mom had been taking me to riding lessons since I was six, where I was encouraged to lead the ponies myself and stay attuned to their body language. Up in the bleachers, I held my breath as steeds swiveled around barrels, running against the clock, exhilarated by the possibility that it could be me whipping through the loose dirt of that arena. At the rodeo, I wasn't just a "Frenchie"; I could be another rider.

My parents separated when I was in third grade. My dad's alcohol abuse—an illness that cycled, destructive and dormant, throughout my parents' marriage—had reached the point of unmanageable. "Daddy's going to live somewhere else from now on," my mom told me as I stood by the front door. She crouched down to meet my eyes and rubbed my back as Tia and Dante wailed on the couch. When the school year ended, my mom packed us up and we spent the entire summer at my grandparent's house in Québec, Canada. My dad didn't come.

That was the year I started riding at a new show jumping barn called Fox Run Ranch in Davie, a few towns over from the Hollywood Rez. My mom rode English, so I wanted to ride English, and once I discovered you could jump a horse, I wanted to do that more than anything; it seemed unreal, like the closest thing to flying. Though the Rez had its own stables, they didn't give English-style lessons, and there were no English categories like hunting or dressage at the Seminole rodeos. As Grandma Hannah realized I wouldn't be barrel racing, our rodeo outings became less regular until she eventually stopped bringing me at all.

Though he was never consistently in my life, my dad's absence from our home was a palpable sore made raw over and over again by Grandma Hannah's dwindling visits and the prying questions from neighbors and classmates. The specifics of this period are a painful blur, an amnesia that, when revisited, allows only feelings of loneliness, immense anxiety, and sadness. A hardening resentment toward my dad took root; I was too young to understand the complexities of his illness and instead interpreted his drinking as a betrayal. How could he do this to *me*? Soon enough, I realized that when I was on a horse, synced to its gait in the arena, my grief became so distant I barely recognized it.

Summers in rural Québec, far away from the humid stench of Hollywood and the fighting between my parents, were an escape into a simpler world. My mom's parents, Grandmaman Jeanine and Grandpapa Gabriel, lived in a trailer-turned-house on 2.5 acres that seemed to extend infinitely into the woods and dirt roads of the surrounding farmland. It was located in a small town called Saint-Isidore, a cool forty-minute drive outside of Québec City. We called it the Canada house.

I can close my eyes and still see every inch of the Canada house and its property. The front porch overlooked a small, man-made pond and waterfall that Grandmaman decorated with flowers and stones collected from the Chaudière River a few miles west. The yard was a cluster of evergreens and birch trees, some with branches swollen around the rope of multiple swings that Grandpapa fastened with sturdy chunks of wood as seats. Behind the house, a stretch of wooden fence contained a pasture of rolling grass and dirt. To the right side of the house, there was a small barn painted white and green. Throughout its days, that barn housed chickens, rabbits, mini goats, and horses.

Despite their years of owning and managing motels in South Florida, neither Grandmaman nor Grandpapa spoke fluent English. This was due in part to the nature of their business, which consisted entirely of Québécois snowbirds, who escaped the harsh winters of Canada en masse to the beaches of Hollywood, Dania Beach, and Fort Lauderdale. Though French was my first language, my mother spoke to me exclusively in English once I started pre-K, and my proficiency never really evolved past that of a child. Simple phrases like "je voudrais" and "est-ce que je peaux" carried my siblings and me through most interactions with our grandparents, which revolved around daily activities like family meals, watching TV, and, most importantly, Grandpapa's horses.

He owned three. There were two retired racing Quarter Horses, the well-behaved Storm and the sweet but sometimes mischievous La Fille Fille. Their sleek, brown coats and elongated features distinguished them as the twin beauties of the barn. But it was Petit Prince, a stocky, temperamental pony with an unruly black mane, who was the star of the show. He was almost half the size of Storm and La Fille Fille, but he was staunchly the alpha. One of Petit Prince's favorite power moves was pushing the poor Quarter Horses aside with bites and kicks to the rear whenever sugary treats were within reach.

Storm, La Fille Fille, and Petit Prince were always Grandpapa's favorite pets, always his most prized. He grew up poor in a small Québécois village where he worked odd jobs and rode local unbroken horses for fun. Horses were a luxury, unobtainable despite his love for them. After years of manual labor and driving semitrucks, he was able to start his own business and move to the States. In his retirement, decades after he first fell in love with horses, he was finally able to possess his own.

The annual summer Blais family reunion was highly anticipated, yet complicated. Much like at gatherings on my Seminole side, I was constantly being introduced to a handful of new cousins I'd somehow never met before. But even more so than on the Rez, my siblings and I were undoubtedly the outsiders. Well-meaning family members would compliment our thick, dark "indien" hair, our "chinois" eyes, how brown our shoulders and cheeks turned in the sun. Once, I overheard an aunt explain to another that the reason I was able to run through the birches barefooted was because of my Seminole blood. Laughing about it was my best defense against the discomfort. The second-best defense was trying not to think about my Nativeness, drawing as little attention to that part of myself as possible. I knew that these people cared for and loved me deeply—that's what made it so hard to identify the wound that their casual othering left behind.

But the unpleasantness dissolved when the evening wound down and my favorite part began: the Petit Prince–drawn carriage. After dinner, Grandpapa would disappear into the barn and, around fifteen minutes later, Petit Prince emerged from its garage door, clacking his hooves on the gravel and towing a little red carriage. If I was lucky, Grandpapa would choose me to sit on his lap and man the long leathery reins that Petit Prince reluctantly obeyed. Year after year, from the time I was four years old to my senior year in high school, every family barbecue or fondue feast on the lawn ended with cake and rides around the property.

Simply put, our childhood in Québec revolved around horses. As the middle sibling, Tia was old enough to accompany me, the oldest, on secret missions into the fenced-in range where Storm, La Fille Fille, and Petit Prince trotted freely. We shadowed their every move like a team of heavy-breathing zoologists. Making sure my mom or Grandmaman couldn't see us from the windows, Tia and I scaled the splintery barricade and worked our way slowly toward the horses so as not to spook them. Their pasture was over an acre, so some days we made a game of standing over manure piles and deep hoof marks in secret search of our specimen. We took turns stroking Storm's flank as he shook off flies and lazily grazed, seeing how far up his tall hide we could reach our fingers.

Eventually, baby Dante was old enough to be left unattended in my care, which meant I brought them along on these daily equine excursions. I soon regretted this, though, the time Petit Prince bucked and galloped aggravated circles around our bare feet in the mud. High-strung and huddled together, the three of us overwhelmed the pony, triggering his brat behavior. Dante cried as we each took turns making a run for the gate under my nervous direction.

As unpredictable as this little herd was, we kept showing up. Not even the next-door neighbor's mares were safe from our outstretched palms and incessant clucking. Every juicy hand-ful of grass and wildflowers, and every apple yanked from the yard's apple tree, became an offering to our four-legged best friends.

When I was seven, my mom sent me to horse camp in Breakeyville, a small town fifteen minutes north of the Can-ada house. There, I was the only English speaker and the only nonwhite person amongst the ten or so campers. It was similar to my experience at the barn I rode at in South Florida, where most of my peers were also white. But in Québec, I grew as a rider. As a kid who was painfully self-conscious of my shitty French accent and what I learned were my obvious "indien" features, the only person I really interacted with was my fa-vorite instructor, Marie.

One morning, Marie and I spent hours in a one-on-one les-son because I was the only advanced camper who showed up that day. Determined to impress her, I repeated her show jump-ing exercises over and over again, sweating into my helmet as she raised the poles, my horse flying higher and higher at my command. When I inevitably had my first major fall and ripped a hole in the crotch of my jeans, we just laughed. "It happen to all of us," she assured me with her thick French Canadian accent and toothy smile. Later, eating limp ham-and-cheese sandwiches with everyone in the tack-turned-lunchroom, I wasn't even embarrassed by my visible underwear; I was proud to have something to show for my dedication.

Marie made me a better rider, but Davie's Fox Run Ranch back in Florida was where I was introduced to a wider eques-trian world. Sprawling and slightly run-down, the land sprouted with green overgrowth while white, chalky sand paved the trails and pens. My siblings and I spent most of our time in the covered, rodeo-size arena where we took group lessons and, during horse camp, painted our ponies' hides with horse-safe glitter and bobbed for apples. Inside jokes were formed with other eager campers and dream ponies were discussed as we groomed our assigned steers from the stable. In Davie, the Blais-Billies weren't unintelligible foreigners—like at school, we were once again those Seminole kids from Hollywood, familiar but still other.

At shows, there was occasionally another rider of color from another barn. Still, I couldn't ignore the chronic suspicion that I didn't belong or deserve to be there. I was competing against girls with purebreds, shiny new saddles, and pristine uniforms—their wealth rolled off their shoulders in the form of innate self-assurance. Even at our barn, I found myself dodging questions from white riders like, "Which neighborhood do you live in?" or, "What does your dad do?" I was never in the mood to explain what a reservation was or that my dad was an alcoholic who disappeared on benders for weeks at a time. We shared a bond over horses, but we never became close. Distance was my friend when it came to protecting myself from shame or rejection.

My priorities shifted at the Canada house as I got older. The year I turned ten, I was expected to help Grandpapa mornings and evenings in the barn, raking the horses' stalls clean and refilling their buckets with cold, metallic well water from the hose. Well into the last year of his life, Grandpapa was as strong and active as someone in his thirties—he single-handedly cared for the animals, chopped firewood for the furnace, and fixed things constantly around the house. On top of our language barrier, he was a man of few words who was hard of hearing, so our barn chores were completed mostly in silence. But talking wasn't necessary; we were doing what needed to be done, hoisting forklifts of hay to hungry muzzles, helping one another take care of the beloved creatures that occupied our days.

It was around this time that my mom allowed me to ride La Fille Fille, who became stubborn and uninterested with a saddle on her back. We took turns warming her up on the lead and then willed her into canter drills around the pasture while Storm and Petit Prince munched on carrots in their stalls. With the increasing chaos of our family life back on the Rez—my parents' separation only seemed to exacerbate my dad's drinking as the years went on—I cherished the semblance of structure and responsibility dictated by the parameters of my quiet, Québécois farm life.

Grandma Hannah likes to joke that my Mikasuki name, Lokaeechete—which roughly translates to "something passing by in the distance"—is a self-fulfilling prophecy. When I graduated from college in Manhattan, I stayed in New York City, opting for the less hectic borough of Brooklyn. Today, I'm still in the same sleepy Brooklyn neighborhood, piecing together my own life far away from the Hollywood Rez. I fly back for holidays, but extended stays are a confusing mixture of triggering and healing. This became especially true when my dad lost his battle with addiction in 2013.

Recently, Dante moved back to the Rez after their college graduation. We talk on the phone often; they've become my closest connection to the place that, for better or for worse, will always be my home. So it's on an unremarkable spring afternoon when Dante calls to tell me about our paternal great-grandfather, Harjo Osceola.

Though we grew up on the Rez, my siblings and I had never heard of Harjo Osceola. We knew about his son, Genesis Osceola, our father's father, but even that knowledge was severely limited. Genesis was absent from my dad's life completely; we never got to meet him before he died.

Since our own dad passed, we've begun the project of reanimating the life he never really shared with us. We cling to every shred of information gleaned through family stories, social media, or Rez gossip. We compare notes, carefully weaving facts and feelings together to fill in the blanks left behind by generational trauma and decades of communication breakdowns. Dante's phone call is energized with the air of potential, a chance to understand who our dad was as a Seminole man, who *we* are as Seminole people.

"Apparently, he was a prominent Seminole cattleman," Dante says. They text me a black-and-white photo from what looks like a history book and explain that their colleague, another Tribal member, sent it to them when she found out who our father was. I zoom in on the photo, a group portrait of seven Seminole men from the 1950s. They sport baggy button-up shirts tucked into heavy denim and lean against a tall wooden fence surrounded by the twisted branches of Florida's live oak trees. They all wear cowboy hats and grip cattle-branding irons. My Harjo's eyes are cast down, his cheekbones small but sharp, his iron spelling "HO."

I wonder if my dad knew his grandfather was a Seminole cattle-man, if that would have made things different between us. Would he, instead of Grandma Hannah, have taken me to the rodeo? Would he have found pride in who he was, giving him the bravery to fight a little harder to stay around? From what I've heard, all it took was one scary fall as a kid in Big Cypress for him to write off horseback riding forever. It strikes me that most of my friends in Brooklyn have no idea that I ever rode. I consult the group Instagram DM I have with my siblings: "You guys ever contend with the fact that we were horse girls?"

Tia—a classic middle child keeping the peace with something witty at hand—answers immediately: "I talk about it nonstop." I laugh, because it's kind of true. When the three of us are together, we're always bringing up our memories of horse camp, of Petit Prince. It astonishes me that I never associated these facts with the broader consciousness of the pervasive "horse girl" memes and stereotypes. "I buried it so deeply in my subconscious," I type, "but I'm finally excavating."

Sure, I was the weird girl in middle school who kept to herself, read horse-themed YA, and sketched wild stallions on ruled paper. But I wasn't a horse girl—I couldn't be. I wasn't a white, rich femme like the Clique's Massie Block or the girls I competed against from places like West Palm Beach and Boca Raton. I couldn't relate to the privilege or sheltered existence that people around me projected onto the young women who openly loved horses.

Even within the barely-there mainstream representation of Native Americans in movies and TV shows, I wasn't the type of Native that non-Natives associated with riding. At my elementary school, where I was the only Indigenous student in my grade, classmates would come up to me during recess and ask me if I lived in a teepee, if I had electricity, if I rode a horse to school. "Um, we have a *car*," I would scoff back to hide the water rimming my eyes. I vividly remember the burn of my face when a kid told me my dad was stupid because he was an Indian who "danced around a fire."

I saw that for my peers, "Native American" evoked only imagery they were familiar with: Plains Indians, like the Lakota and Blackfoot, frozen in the nineteenth century. To them, I was buckskin, headdresses, and Appaloosas with war paint. That's just not what my family looked like. Our history classes also taught them that Natives were dangerous "savages" who were vanquished by our forefathers because they were intellectually inferior. We learned that the first "American Dream" was "Manifest Destiny," the delusion, veiled as divine purpose, that Christian settlers were destined by God to expand across the New World. I've lost count of how many times children and adults alike have said, "I didn't know you still existed" to my face. Off the Rez, I was either invisible or an uncivilized relic on horseback.

By the time I started high school, my parents had been separated for around seven years, and my dad's drinking was at its worst. He became a ghost of the person I loved, so I rejected him and the parts of myself that were him, alchemizing my broken heart into anger and self-hate. I wanted to be all

Frenchie and no Indien. I wanted so desperately to be like everyone else at the small, predominantly white private school I transferred to and attended on a scholarship. I wanted to be thought of as "normal," and maybe, one day, even *cool*.

I quickly surmised that horseback riding would not help my case—not as the only Native American in the class, and most definitely not as a socially awkward newcomer who still shared a bedroom with their sibling because our HUD house was so small. I went from daily lessons to biweekly to none at all; competitions became a thing of the past. I felt guilt when Grandpapa asked me about riding once I'd stopped. But who could blame me for abandoning a world that never fully welcomed me to begin with?

What I've been taught since I can remember and what I know is true: the Seminole Tribe of Florida never signed a peace treaty, never surrendered our land or our people to the United States. In college, studying the nuanced Indigenous history I wasn't exposed to in the Florida education system changed my life. It gifted me with the vocabulary to unpack my experiences as an Indigenous woman and a space for me to contextualize the pain my dad experienced as an Indigenous man, how he coped with what was available to him. In his death, I forgave him, accepted him, and began to accept myself, too.

But when I follow the Seminole Tribe of Florida's history with cattle and horses, our unconquered resilience goes back even further than I imagined. According to leading Seminole historian and anthropologist Patricia Riles Wickman, the mastery of cattle herding and horseback riding defines the Tribe's relationship with colonization. It was the ancestors of the Seminole people—the precontact tribes of what is now known as Florida—that were some of the first Indigenous nations to husband these animals.

In the late 1500s, Spaniards established the first permanent European settlements in what they called "La Florida." The land was occupied by many different societies, including the Maskókî, Hitchiti, Calusa, Yamásî, Chicása, Apalachi, and Timucua tribes. As foreign invaders, the Spaniards' dealings with the original peoples were an ever-evolving hybrid of violence and diplomacy, depending on what they needed.

Much like the rest of colonial history, the Spanish established mission villages to "save" the Indigenous peoples' souls and assimilate them to Western ideals of civilization through the vehicle of religion. (Because of this, my dad instilled in us a healthy dosage of skepticism when it came to Christianity; I still reflexively cringe at the mention of Jesus.)

The Spanish also incentivized the Natives to work cattle "ranchos" with land grants, which, of course, consisted of land the settlers had previously stolen from said Natives. With a plot of land and some livestock, the Natives worked raising and selling animals back to the Spaniards at whatever price they were willing to pay. This is how they exploited Indigenous labor for profit. By the late 1600s, the Maskókî people established high-profiting cattle herds in what's become today's Alachua savannah in Micanopy, Florida; this prairie is where the group that became known as the Seminole people began.

Throughout the five hundred years of violent colonization and systematic genocide, the Indigenous peoples of Turtle Island (North America) always rebelled. In La Florida, many were subjugated, but many also fought and escaped, finding refuge around the Alachua savannah cattle rancho. There, peoples from various tribes across the occupied land were able to find work and thrive outside of Spanish, then British, and then US control. Natives from territories neighboring Florida, as well as escaped African slaves, were also moving south in search of freedom.

Using the term their countrymen coined to describe escaped slaves in the Caribbean, the Spaniards called this growing, diverse community of rebels "cimarrones," which meant "wild ones" or "runaways." After decades of playing telephone through multiple Indigenous and European languages alike, "cimarrones" became "Siminolie" and then eventually, "Seminole."

It's satisfying to see how poorly the Spaniards' master plan played out; I smile to myself knowing that my ancestors one-upped their oppressors. When the Maskókî and Apalachi and Timucua and Calusa and Hitchiti and Yamásî and Chicása people were forced into labor, they adapted. They learned the ins and outs of ranching, applying thousands of years' worth of knowledge about the land to the craft of keeping herds alive as the runaway Seminole peoples. They studied the economic customs of the settlers, becoming acute businessmen for the survival of their communities in an ongoing war.

Even hundreds of years later, during the bloodiest years of Seminole resistance against the US, the skillful husbandry of scrub cattle and horses kept the remaining few hundred rebels fed and alive. The Seminole Wars, three in total, spanned from 1817 to 1858 in a concentrated effort to eradicate the "Indian problem" of Florida. Initiated by the notoriously belligerent Andrew Jackson, the series of concentrated military efforts failed to completely remove the Seminole people every single time. During the Second Seminole War alone, the US spent almost $40 million to try to capture and relocate around 3,000 men, women, and children to Oklahoma, or "Indian Territory." It was also the only Indian war in American history that employed the army, navy, and marine corps—unsuccessfully, I might add.

I find Harjo Osceola listed in the United States Federal Census for Hendry County, Florida, on Ancestry.com. The document states he was born in 1913 and died in 1978 at age sixty-five. He lived twenty-six years longer than my dad.

In a 1972 interview with the University of Florida's Samuel Proctor Oral History Program, Harjo's younger brother, Reverend Billy Osceola, detailed the family's life in the early days of the Seminole Brighton Reservation near Lake Okeechobee. "That time, we [didn't] have any reservation," he stated. Harjo, Billy, and their siblings harvested their father's vegetables down in Big Cypress to feed themselves. They hunted alligators, raccoons, and otters, selling their hides and meats to white traders in Indiantown. Their mother, Nancy Osceola, passed away when they were just teenagers.

They spoke English and Muskogee, or Creek, which is one of the main languages Seminole Indians still speak today. My

dad spoke Mikasuki because for a good portion of his childhood he was raised by his maternal grandmother, who didn't
speak a word of English. Grandma Hannah attempted to teach
me some words; sadly, I can count the amount of things I
know how to say on one hand.

Googling "Seminole cowboys" in my Brooklyn apartment,
I try to envision Harjo in the tanned faces of Tribal men
staring indifferently toward the camera on their steeds. By
1957, when Harjo wrangled the fields, the Seminole Tribe of
Florida became a federally recognized tribe. That meant the
Brighton Reservation was officially an agricultural and livestock enterprise independent of the US government. The Tribe
was able to appoint their own land trustees internally, eventually using that legal foundation for future land claims and
reparations. I write this down, feverishly, in anticipation of
telling Dante that our relative's work was crucial to the political advancement of our Tribe. Our great-grandpa!

Researching Harjo reminds me of one sticky night on the
Rez. I had just moved to Brooklyn and I was home decompress-
ing for a week. It had been three years since my dad died,
and Grandma Hannah made a habit of coming over with her
tóhche, who we called Uncle Paul, to tell us stories in the backyard while the sun sunk behind the freeway. Mosquitos pricked
our backs and cicadas whirred, but Tia, Dante, and I listened
attentively. Uncle Paul spoke about the way-back-when times
of the Seminole Wars, detailing the lore and lessons our people bore in hopes of one day telling their great-grandchildren.
I eyed the darkening skyline, my heart swelling tight at the
image of my ancestors slithering into empty alligator burrows

to evade US troops. "They knew the land like no white man did," he said.

Tia and I have settled on the loose term "Seminole horse girl." It seems simple, but the specificity allows just enough space for the intricacies of our biracial identity. Like the Seminole peoples, "Seminole horse girls" originates from a conglomeration of cultures adapting to their environment; sometimes not belonging to one group exclusively can be empowering. I've found that, in our family, horseback riding is more than show titles and prestigious stables—horses are how we survive.

For so long, horses were the love language that kept me tethered to the peace and stability my grandparents offered by way of sharing their home every summer. The Canada house was a place where I was safe to explore, process, and build the confidence I wasn't able to in the noise and identity politics of South Florida. All those years of horseback-riding lessons, horse camp seasons, new riding boots and tack—those were reassurances from my mom that we were just as good as the snooty kids at competitions, that we'd have fun and meaningful lives, no matter how unstable our household.

A few months after my dad died, Grandpapa passed away from cancer. The year before, in 2012, he had to put down Petit Prince due to an incurable abscess in his mouth that caused him to stop eating; Storm had already succumbed to lung disease a couple years before that. Seeing Petit Prince's death as a sign, Grandpapa sold his remaining horse, La Fille Fille, to "une gentille vieille dame" who lived not too far from Saint-Isidore. It was a hard decision, but he visited her a couple of times in her new home. She seemed happy. Less than eight months after Grandpapa's passing, we said goodbye to Grandmaman, too.

I think about my Seminole ancestors every day, like I think about my dad and my grandpapa and my grandmaman. So much was lost, and yet, I am here now. I give thanks for what they've done and who they were. I want to stop strangers on the street, grab their shoulders and shake them and scream, "Do you know what they meant to me and to my understanding of myself?!" Though my Seminole and French-Canadian sides feel like worlds apart, they each gifted me with a reverence for horses that bridges the distance that once overwhelmed me. I follow the bloodline down and I see who I am: July, Genesis and Grandma Hannah, Harjo; France, Grandmaman, Grandpapa. I ride bareback in a sunny Canadian field and wonder, if Harjo could see forward, would he see me? Certainly this equestrian dynasty, this way of living that connects me to all my loved ones, could warrant such foresight.

FOR THE ROSES
ALLIE ROWBOTTOM

The day of the show begins in silence, the fairgrounds still before the sun comes up, the arena closed, the barns dark. I stand on a stool beside the horse. His head bobs in sleep, my stomach presses his neck. Lamps clamped to the corners of the prep-stall warm my work, the braids I make out of mane and tail and sticky hairspray, the black yarn I add in to tie each plait.

I finish and the announcer's voice starts up, echoing down barn aisles—*testing one two*—to the far edges of the grounds, where it fades. I dress myself, saddle the horse. "It's time, Hammy," I say and unhook his chains, offer a bit, cold and long-shanked, shaped like a weapon. He opens and takes it like he wants to. *Ham*, I say and tighten straps, straighten reins, *Ham Ham Ham*, a mantra.

Ham is the horse's barn name, his nickname. I gave it to him when we were young and our story was beginning. Before me, he went by Rambo and seemed dedicated to becoming his namesake. He pawed trenches in his stalls, pinned his ears, bit.

"You're such a ham," I said the day we met, to convince us both he could be different. With me he wouldn't need violence. I would be his girl, and my love would transform us.

Now, five years since Ham came into my life, and I am

nearly a woman. I am eighteen, about to age out of the junior exhibitor division in which I've competed for half my life. It is the morning of my last ride, my last chance to win the World Championship title I've wanted my whole girlhood. I try not to think about it, the ending, the last chance, the potential for loss.

I mount and my body becomes Ham's. We walk from the barn and into the morning's music: metal shoes on cold pavement, the crack of a whip, the jangle of chains, wrapped around a show pony's ankles, to make him trot high and square and rhythmic as a drumbeat, calling me.

The paved strip between small show arenas leads us toward the Coliseum, where World Championship titles are vied for, awarded, lost. As we near, I visualize our ride, picture each transition, each pass before each judge. Visualization is a technique my father taught me. Through it I tap into what he calls *pure Force*; I become one with the universe, limitless in my power.

The Force is a concept my father learned from his favorite book, which shares a title with its signature concept. In *The Force*, author Stuart Wilde, seemingly unaware of the *Star Wars* entendre, teaches of a "massive, exhilarating, magnanimous" energy in and around all things. Only those who believe in the Force can tap into its power. But once we're tapped in, we can visualize anything we desire, and it will manifest. I visualize Ham, ears pointed forward, eyes bright; I see him speed up and slow down when I ask him to; I feel him breathe, I watch us win.

A golf cart carrying a group of girls whines up behind us and returns me to the moment. They speed past, tan legs hanging. *Good luck*, I hear. They shrink into the distance. Ham and I steer into the A Barn practice ring; we move from bright morning into a dull, familiar light.

I know to expect the dark. Once a year for the last five years, I have taken this path, moved from this kind of day into this same muted space and then, the Coliseum. Each year I have ridden to win a World Championship in Junior Exhibitor Hunter Pleasure, a division made up of girls under eighteen years old, and their Morgan horses, all of us outfitted for a hunt we'll never ride. We ride, instead, in circles, performing the commands of a disembodied announcer: *walk please, ladies, come down to the walk; reverse directions at the trot; canter please, riders, canter please.* The most beautiful pair to follow these commands without fault wins.

Each year I've ridden for the World title, my father has stood on the rail, arms crossed, eyes trained on my body and Ham's, like he might direct us with his gaze. Each year we have been perfect, the most beautiful. We have been perfect until I've made an essential error, a rookie mistake. It is something about the World, I sometimes think, the pressure, the heft with which I want to win; it is something about the wanting that makes me choke the way I do. Every year we've tried for the roses, I've let other horses envelop us, box us in, and Ham has reared up; we've been subject to surprise attacks, riders I didn't watch for, cutting us off, shaking us from our strategy. Ham has thrown leads, flipped the bit, taken the metal shank in his mouth, and wrenched. Which is to say, for six years we've traveled sixteen hundred miles, from New Hampshire to Oklahoma City, and lost.

A long, downhill chute connects the warm-up ring and Coliseum. Already, horses and riders—horses and girls—outfitted and groomed like us, have formed a line. Ham and I take our place among them. As one we wait, soldiers on the front, uniformed and ready. I shorten my reins, glance around to see who might be watching, who might be afraid. Ham is not the only beautiful horse and I am not the only girl riding her last ride. The World Championship will be mine, or it won't. And regardless of the outcome, for the first time in six years, my father isn't here to see.

Come on in, ladies, the announcer calls. The first rider descends, the second follows. They pass into the Coliseum and the stands erupt with cheers, shouts, whistles. Ham and I walk down the chute. We near the gate and begin a trot.

There is a path before us, so bright it blinds. *Riding for the World Championship*, the announcer booms. I push us into the humming light.

I am eight years old. Stirrups too short, helmet askew, tufts of curls puffed out around its edges and tangled in the straps. I ride through the early morning fog that hangs over the cornfield and river and dusty ring, too big for just one girl. I trot it anyway, measure circles with gentle geldings my teacher, Andrea, assigns. I check diagonals and leads, practice equitation, back arched, thumbs pointed to the sky, a doll in the saddle.

"Heels down, Al," Andrea yells. She is tough, calloused, her voice nasal and blunt. She stands center-ring, head shrouded by the hood of her windbreaker, coffee mug abandoned on a rusted folding chair.

"You can come in here, George," she calls to my dad, stationed on the fence-line.

"No," he says, "I'm good." He prefers the rail, the outer edge; he waits there, and studies.

When the lesson ends, Andrea returns to cleaning stalls and my father returns to the car. I ride around the cornfield, past the river and back toward the circle, as if surveying the bounds of my world, invisible walls beyond which I am not safe. My horse points his ears; he scans the distance. When dry stalks rustle and let loose a flock of blackbirds, we both jump, then settle.

Horses smell fear, a common lesson for young riders; bury your fear, and your horse will, too. Maybe there's truth to it. Horses and humans are ancestors, after all. Hundreds of millions of years ago we crawled together from the sea. From a single-cell organism, a mutual mother, we diverged. Skeletal vestiges prove our bond: identical patellas, for example; or the modern horse's hock joint, akin to a small bone in the human foot. So perhaps we share survival mechanisms, too, subconscious memories: how it feels to be abused; what it takes to live through war.

I am eight when I begin to ride, nine when I decide to win. My parents and I live in a log house perched high on a hillside. It's a cavernous space, and cold; the echoes of anger bounce from every wall. They fight most in the kitchen, the counter-top between them. I climb up and lie down on the butcher block, that neutral space. I demand each person stop. I am a peacemaker, but the mission is one I always fail, abandon.

For a time, I retreat to the woods, but they're thick, all forest, no paths. So I beg to ride, search the phone book and find the farm, sign up for nightly chores, morning lessons. Five a.m. and I'm waiting in the car, ready to return. My father climbs in the driver's seat with bowls of oatmeal for us both. We roll down the rutted driveway, spooning in bland bites, and it feels like an escape.

Before the woods, my parents and I lived by the sea. My mother was sick and my father was scared, angry when he couldn't help her, as if cancer was a failure of his will, or hers. I was an amateur whale enthusiast and planned an adulthood in marine biology. Now, the forest surrounds us and my mother is in remission. But my father's rage persists. Now, it's horses I love. I pore over easy-reader young-rider books, to learn their history, from war mounts and agricultural aids, to saddle clubs and shows. Andrea gives me her outdated copies of *The Morgan Connection* and *The Saddle Horse Report* and I memorize show results, learn the foundational bloodlines of the Morgan breed, to which the farm herd belong. I learn the rules of showmanship, equitation, proper grooming and care. I learn the other farm girls, learn who works hard, who slacks, who Andrea thinks is aggressive enough to win, who she thinks too soft.

Soft or not, on summer afternoons she teaches all of us equine anatomy on Spunky the one-eyed pony; Sundays, we trailer to 4-H shows, earn ribbons, mugs of candy; we return sunburned, and full of glory.

My father hangs photographs of my victories on the foyer wall. It's a drafty, transitional room, full of other pictures: my father's parents, whom I never met; shots of his own eighteen-year-old face, dirt streaked and beautiful behind the wheel of a car. And a few black-and-white photos of a boy my age. Here he stands and grins, haloed by sunlight. Here he is somber and holds a dead bird in one hand, the gun that killed it in the other.

"Who is this?" I one day think to ask. I point at the boy.

"That's me," my dad says. He sounds surprised that I don't recognize him.

The longer we live in New Hampshire, the longer my mother stays healthy, the less I recognize my father. He spends his time alone, locked in his office, the room overlooking the drive-way from which he tries to manage everything—comings and goings, finances and stock futures, mealtimes and homework schedules, bedtime and morning rituals.

Everything in the office once belonged to the grandfather who died when I was three: the long wooden desk and roll-ing chair; the tattered green recliner where my dad meditates; the racks of antique firearms my grandfather collected, each one responsible for the dozens of taxidermied animal parts he also bequeathed to my dad, so many that they fill not only the office, but the entire house. Busts of impala and gazelle hang from every wall. Elephant legs serve as end tables; the feet of wild boars, severed above the ankle and topped with metal ashtrays, adorn every coffee table. The head of a Cape buffalo, taller than a man, hangs above the fireplace; the skin of a lion, its head stuffed and still intact, its mouth gaping, stretches across the floor, shedding strands of wiry mane. In the photographs of my grandfather that hang in the foyer, he is gun-slung, fragile-bodied. He kneels next to the animals he's hunted and killed, his palms on their bodies like a healer.

I am used to these items, these trophy animals. They feel like protectors. My mother complains they feel, instead, like artifacts, the house a monument to power, wealth, white masculinity, so much dangerous privilege. She makes art in protest, dioramas of animal figurines: plastic elephants, whales, horses, sawed up and reassembled in surrealist jumbles she places around the house, side by side with my grandfather's kills.

"A total clusterfuck, the things I've been through, the things I've had to put up with, and never say a fucking word about." This is what my dad says when we speak of why his dream career failed to launch. He had wanted to be a professional race car driver. "You would be blown away if you knew what I've been through, but you never will, because I'm not that kind of person. I don't dwell." He talks to me like I'm a grown-up, capable of relating. I nod along, listen like I think a grown-up should. "I had the talent," he says, "I had the ambition. I had everything but support. My father didn't help me the way I help you."

My grandfather was larger-than-life, charismatic. "He was a genius," my father says, and tells me that during the "Great

Wars" he built propeller parts and firearms, then traveled from place to place, teaching soldiers how to use them. But he was also haunted: alcoholism, a sense of having failed to make more of what his parents left. It's a feeling my father inherited, too, a chromosomal curse.

"Whatever it takes," my dad says of my riding, the World Championship title I hope to someday win. "I'll do whatever it takes to make your dream come true." He will leverage what's left of his father's wealth. He will play the stock market to earn enough for a horse, a prestigious training barn, a private plane that he himself will fly, all the way to Oklahoma City, the World Championships, where I will win and win and win.

"Picture this," he says. "You're a World Champion, then next thing you know, you're a professional trainer. You wake up in a mansion in Northern California. You look out the window and see a field of horses, all for you. You go to the basement, and find barrels of wine."

I prepare for my future, belly down on my bedroom rug, studying horses, their history, a long lineage of winners and losers into which I must fit. I learn that scientists date horses' domestication from approximately 5,500 years ago, but quibble over the advent of riding. Most believe that by the year 1500 BCE, horses were pulling chariots, and by 900 BCE, humans were riding them. These dates mark the beginning of a global tradition in which horses were commodified, turned to tools of conquest, tools of men.

For centuries, regardless of location, conflicts between cultures with different approaches to horses boiled down to a simple fact: winners rode. An army on horseback always trumped an army on foot. But because horses were also often symbols of aristocracy and priced accordingly, they were typically in short supply.

Even Napoleon Bonaparte, whose fearlessness and drive to win has become legend, confined his cavalry to noblemen and at times struggled to drum up enough animals to supply his officers. Napoleon's own horse, Marengo—the rearing white stallion whose image is inextricable from his owner's—carried Napoleon to numerous victories. Until, that is, the Battle of Waterloo, which Napoleon fled, leaving Marengo wounded on the road.

Horses were conduits for Napoleon's victories, bodies he treated as extensions of his own until they failed and he moved on, saving himself. This, I suppose, could be considered the personality of a winner. But I have seen it become a formula for loss.

After my early morning rides, my dad drives me to the school bus stop. When we miss the bus, as often happens, he parks, slams the steering wheel with both hands. Then he stills and stares into dead morning air, the town's single stoplight, the single Sunoco station beyond it. He makes a shadow-puppet shape with his right hand, lifts it to his nose, blocks one nostril, and inhales. He blocks the other, and exhales.

He drives me to school, still breathing this way, thirty minutes to the soundtrack of his snot, the stuck, negative emotional energy he needs to cleanse. "Cleanse" is a word my father favors. To tap into the Force, he tells me, we must first cleanse ourselves of bad feelings, trauma, fear. This is because the Force mirrors our thoughts and feelings; if we feel joy, happiness, success, the Force will deliver more joy, happiness, success. If we feel afraid, the Force will give us more to fear. Fear is toxic, the result of toxic friends and family, the toxic past we must release through isolation, positive affirmation, visualization and strategy, expedient silencing of negative vibes.

"Cut that out," my father says when I cry, as if my tears might poison us both.

"Good job, Sport," he says when I win my walk trot classes, as I often do.

Sport was what his dad called him, a nickname I heard for the first time the first time I won. Sport is who my father wants me to be, a child in the image of the champion my father should have been, a child who always wins and never loses, a child I secretly fear I'll never be.

"We're the same, Sport," my dad tells me when I win. "We're different than other people, I understand you."

"Jesus Christ, Allie, only *you* can make yourself feel bad," he says when I'm myself, feeling too much, wanting too much: food, attention, affection, all of it excessive, unbecoming of Sport.

"Make it a great day," he says when he drops me at school. I attend a private academy in Vermont, a river away from the farm, and my report cards reflect the distance. A moony child, they say, a girl whose mind is somewhere else.

"I will," I promise, and get out, slam the door harder than I mean to. I know my father's morning, the early hours after the stock market's bell, have been lost by the missed bus, the long drive. Most likely he will still be angry about it when night comes and he returns me to the farm for chores.

I sit through science and math, subjects I am slow to understand. I sit through American history, search for myself in stories of men and war. I tuck the collar of my sweatshirt over my nose, my mouth, and inhale. I smell like sweet feed and horse sweat, a certain wildness. It comforts me.

My history books are full of horses, so I like the subject best. Grant's Cincinnati, Stonewall Jackson's Little Sorrel, and Lee's Traveller, they hover in squares on every flimsy page; small blocks of text tell their stories. I learn that during the American Civil War, even as both armies struggled to obtain enough horses, they trumpeted the supernatural bravery of their cavalry. Tales of special horses, horses with transcendent powers, were tossed back and forth across enemy lines like cannon fire. Certain horses could, soldiers claimed, intuit the arrival of enemy forces long before their riders; should a rider fall off or die on horseback, his mount returned to her home army, her band of fellow warhorses, of her own volition. She didn't need her freedom, the story went, her loyalty trumped it.

The reward for loyalty was dignified death, a place in history, a bronzed statue in the town square, a framed photo, memorialized on the foyer wall.

I watch my body change in photographs of victory passes, show seasons, levels of competition. Here where I'm a 4-H rider, dimpled and small, a winner of candy and hollow plastic trophies. Here where I've graduated to A-rated shows but my hunt coat puckers and my last-place ribbon is purple, brown. Here where I am taller, thinner, almost adolescent, and back in the red and blue, back to being Sport, but burdened by my body's power, failure, fear.

Fear is toxic, my dad warns. Horses smell fear, the Force mirrors fear, enemies feed off it; he doesn't want his girl to be afraid. But when a bearded man at the General Store—an artist, he says—asks to take my picture and I agree, my dad erupts in rage. I should know to fear that man. I should know I am a girl, not a horse, not stronger than a man, not faster. Fear is something I should know to pick and choose.

I cancel the photo shoot, start high school at a new academy where I know no one. My parents give me a cell phone, and a group of popular boys gets the number. I wake in the middle of the night to messages from unknown callers. "This is the vet, it's about your horse," a gruff voice says, and I panic. "I've heard you like horse cock, big hard horse cock." In the background, laughter. "I think you should try my big hard horse cock."

The messages are my secret; they are mine to be ashamed of and I fear their arrival. When they come, I press delete, lie awake and hope they don't return. At school, I trudge around, shoulders slumped in, like my body is an overcoat, pockets full of stones. I fantasize invisibility, which only riding offers. On a horse I am not a girl, not a body. I am a horse, and closer to escape.

Escape is a privilege for any girl. Protection is a privilege. My father makes this clear. He sits in the driver's seat of his parked car and pulls his checkbook from between the seats. He writes out numbers so I can see how much he pays for my lessons. He signs his name in an anxious scribble, tears out a slip of blue, hands it to me to deliver.

A particular ritual, a weekly ritual and extra costly when my parents finally separate. They refuse to speak, let alone fight, and hire lawyers to do the talking, screaming, bargaining, for them. Around me they are guilt ridden, worried. They move through separate fogs of grief and anger, and money flies from their hands like it's abstract and meaningless. They drain my college fund to buy me Ham and pay his board at a fancy training barn on the New Hampshire seacoast. They take turns driving me two hours west each weekend, so I can practice for the coming season, my graduation from the "thirteen and under" junior exhibitor age group, to the "fourteen through eighteen" division in which Ham and I will show.

When we meet, Ham is called Rambo. He is violent, but with time, attention, a different name, he begins to soften. Still, my new trainer keeps all her horses in stalls; they never go outside, and Ham is no exception. His isolation is for his safety, she says. Outside, Ham might spook, run, hurt himself. "We can't trust him," she says, and smiles.

Ham is not, it turns out, closer to escape than I. We are both isolated, both fearful. He lives in a cage, and I am a friendless freshman. My parents are at war. For the sake of my show career, I have left the farm and ride now only on the weekends.

Without barn chores at Andrea's, the weeks are long. I spend my free time locked in my room at either house, waiting

for Saturday's long drive to Ham, and memorizing show results in the *Morgan Connection* magazine. I note which horses
and girls to dismiss, which ones to beat; I finally read *The
Force*. The book makes sense to me. The book teaches that I
have the power to effect great change. I already believe this
to be true. My mother's illness, for example, onset in tandem
with my birth; my father's anger, my parent's divorce, which
I feared and so, by the logic of *The Force*, created. "You create your own reality," *The Force* reminds me, and everything
that's wrong in my life began as a bad feeling I dwelled on, and
manifested. To do better, I must think only positive thoughts,
feel only positive feelings; I must distance myself from people
who might doubt my ability to win or, worse, compete with it.

"Don't you want friends?" my mother says. She stands in
the kitchen of her new house, hands on her hips.

"I have friends," I snap.

Once she and I were bonded. We spent time together,
grateful that we had it. She had been sick, and I had feared her
death. Now, *The Force* has taught me not to fear. Or divorce has
changed what scares me. Since my mother left, my father has
weakened, his rage and grief so drastic, I wonder what they'll
create. When I'm at my mom's, I lie awake and imagine finding
my dad's mottled body, hung by his own hand, swinging from
a beam in his empty house. The vision rises up and I quickly
tuck it away, somewhere in my body even I can't see.

"Where are you?" my mother asks. I say nothing and she
answers herself. "You're shut down," she says, "always somewhere else." I am in another world. I am at school, the barn,
my dad's house.

"Whose fault is that?" I spit, like she should have stayed.

I know she's better off here, in this new house, sweet and squat, with a wood-burning stove, maple syrup pails in the yard, a rain barrel. My dad's cavern on the hill was always too big. Now he closes off whole rooms to save heat, lets toilets dry up and rust. In the summer, black flies crawl from between the logs of his office walls. Ladybugs infest my mother's kitchen. I turn fifteen and my parents, unwilling to call each other and coordinate, give me the same presents; when I unwrap a duplicate book, sweater, set of stubby silver spurs, at my dad's house, I feign surprise. All I want is to win anyway. And I do, for now. For now, Ham and I coast through the show season as undefeated champions, projected to win the World Championship in October.

My father calls me Sport and helps me visualize my rides. We sit at the kitchen counter, that old battlefield, and close our eyes. I picture Ham, bathed in roses, a spotlight; I picture my body, blotted by the glow.

As Worlds approach, school starts. Since my parents split, since I quit barn chores at the farm and read *The Force*, my grades are up and I'm in accelerated classes, studying European History, doodling horses in textbook margins, careful to erase them after. I listen as I draw; I learn that despite propeller parts and artillery, the likes of which my grandfather and his father produced, during the Great Wars European armies clung to cavalry-specific strategies, and the death toll, both human and equine, staggered. In cities and on farms, horses were still effective tools of labor and transport, but they were wasteful—in 1900 New York City, horses produced 1,100 tons of manure a day—and prone to spooking and bolting. The automobile, high-speed trains, motorized tractors, were rational alternatives to so much excessive waste, excessive instinct, need.

But with technological solutions, critics complained, came a loss of connection, the enmeshment of human and nature exemplified by the enmeshment of human and horse. Machines, it was argued, sucked men's souls, but horses fed them.

First Wave feminists waded into the debate. With a vehemence cringe-inducing to subsequent feminist movements, first wavers relished the old Aristotelian ideas of women's innate connection with anything animal, physical, emotional. As they campaigned for freedom from so-called chattel marriages, they also advocated for the rights of animals considered chattel. And so the corset, the sidesaddle, the layers of heavy fabric designed to hide women's dangerous flesh, fell from favor alongside the "bearing rein," a tight strap used to hold carriage horses' heads at painful, unnatural, and life-shortening angles.

Time passed, technology advanced; women left the home in greater numbers; horses disappeared from urban landscapes and farms, became objects, pets, hobbies, a perceptive shift that primed them to belong, in the Western cultural imagination, to women. Which is how we arrive at a modern day in which Western women and horses are often framed as uniquely bonded. "It is more than a relationship, more than an attraction. The thing is undeniable, even indescribable," writes GaWaNi Pony Boy, editor of *Of Women and Horses*, an anthology both parents gave me on my fifteenth birthday. "Do women have an innate gift that allows them to commune with our equine brothers and sisters?" Pony Boy asks.

For many women, the answer is yes, perhaps because, like it or not, our bodies define what we're considered capable of. Embodiment is a burden, a history, we can't escape; horses, valued or discarded for their bodies depending on situation and cultural mood, remind us of ourselves.

Innate or not, many women riders perform our bond with horses in the competitive sphere, reenacting equestrian pastimes: the fox hunt, the battlefield, the cattle drive. Of course we do. Why wouldn't women want to reimagine a history that has excluded us? Why wouldn't we want to align ourselves with "masculine" traits, or insert ourselves into masculine spheres? Why wouldn't we want to ride the wild warhorses, become fighters, hunters, cowboys, explorers? *On a horse I am not a girl.*

It's October, the week before our first World Championship show, our first shot at the title, and hives erupt from Ham's body like a badly buried secret. He trailers to Oklahoma anyway, shivering with itch, tied to a hay bag he won't eat. My father says, "Use the Force to visualize the welts away." But mantras, oatmeal baths, antihistamines all fail to erase their prickle and stick.

We decide to show anyway. Stake night, my dad sits on a trunk outside Ham's stall with his eyes closed. "I see it," he says, and tells me about the Winner's Circle, the roses, all that glow. In the warm-up he stands on tiptoes to whisper in Ham's ear. The only word I hear is "light."

I try to see the light my father sees; as Ham and I descend the chute, I tell myself it's mine. I touch my spurs to Ham's lumpy coat, ride deep in the corners and aggressive on the straightaways. The three judges see us. My father does, too; every pass I make I watch for his body, grim and focused, beyond the out-gate. He's tapped into the Force, guiding us with the force of his feelings, but it's as if Ham's trotting through water, every step weighed down, and no matter how I ask for more energy and attention, he can't deliver.

When finally the work is done, the judges turn in their cards and all the riders retire to wait. Minutes pass. The judges deliberate, the crowd stirs. *The World Champion is*—the announcer calls a number, a name. The audience cheers for a horse, a girl, who isn't me. This happens for the reserve champion, too. Third place, fourth. In the end, Ham and I leave the ring without a ribbon.

I know not to cry about it. My dad holds Ham and I administer an oatmeal bath. After, he takes me to a chain restaurant and I shovel chips into my mouth, manic in my attention to their sharp edges and caloric worth. My father watches, wordless. He doesn't like to see me lose control. Sport would never lose control, get too fat, too thin, too emotional about defeat. The check comes and he scrutinizes the sum total, sighs like it's too high. Eventually, he pays. We return together to the long New England winter.

A season of loss. My own, my father's, the money his father left him, gone on the day Enron crashes. My father screams like he's been shot. He breaks a framed photograph of Ham and me. He sobs, "What's wrong with me?" The Force, Stuart Wilde, his daughter, the dog, none of us can answer.

Time passes. One by one, my dad removes his father's guns from their racks and sells them. The heat goes off, returns. Taxidermy trickles away. The dog dies. Soon, only the two of us remain.

Summer comes, then another. Bills pile up on my father's desk. I work as a groom to pay them. I groom more than I ride. I travel between temporary barns and braid other people's horses. I feed myself in one hundred–calorie doses, keep count in my head. Just as my body begins to turn toward womanhood, I starve it small again, starve it back into a girl.

"I got it, Sport," my dad says with every bill. "Trust me." He says this as the debt keeps growing and the phone keeps ringing, collection agencies and credit card companies jamming the line until my father unplugs it. What he does for money, I no longer know. Nor do I ask. Nor do I trust him, though I feel his pain like it's my own. I work to turn off the pain, work to save us. I work to become the girl my father wants, thin and willing, a Sport, a soldier for the Force, a soldier in the war against fear.

But in the show ring I lose like it's my job: regional titles, national titles, world titles, the trust of my horse, who I no longer have time for. Only my father still believes in us. "I understand you," he reminds me. "We're the same, you and I." And he's right, he does understand. My mother does not.

As my last season in junior exhibitor approaches she wants to sell Ham. I have college coming up, she says. And my father isn't keeping his end of the financial bargain; she's tired of shouldering more than her half of Ham's expenses, and mine.

"You don't understand," I scream at her, and we fight, always returning to the fact of my failure. Because no matter what I earn, it's never enough to cover my father's debt, the mountains of delinquent bills he hides from, the running list of what he's supposed to split with my mom and me, but never can.

Money is a problem for everyone but him. "There's no such word as *can't*," he says. "Trust me." He finds me a car I pay for. He gives me a credit card for gas. It never works, and I steal from my mom's purse to make it to the foot of his driveway, where I leave the empty car, walk the rest of the way, up ruts that run like veins to the cold house, the unplugged phones, the missing guns, my father in his meditation chair, manifesting.

"He's in a fantasy world of denial," my mother says. "You're both in a fantasy, pretending you're rich."

Time passes, and I begin to suspect she's right: I'm not Sport and my father's not pure Force. But I'm still with him in the dream that he is. I'm with him in the dream and I'm at war with who I have to be, to stay there.

I do my best to heal. Weekends, I drive to the barn, visualizing freedom for myself and Ham, a grassy field in which he grazes, a college I attend, far from my parents and their pain.

At the barn, I exercise Ham before I get to work grooming, lunging, stall picking. I put him first and go slow when I ride, reminding myself to soften the way I want my horse to soften. I speak to him, say his name, tell him he's safe, there's nothing to fear. After we train, we ride outside; I loosen the reins, stand in the stirrups, let him run.

It's then I develop a new obsession, a secret history I study to rewrite my own. In library books and novels, on the internet and in myth, I find horsewomen. I find them painting mustangs on cave walls and milking mares on Mongolian steppes. I find them in the Celtic goddess Epona, a woman, a warrior, a horse. And in the Amazons, empowered by their rage. I find, in Russia, Catherine the Great, who refused the customary sidesaddle and demanded that the women of her court follow suit. In a famous portrait by Vigilius Eriksen, she sits atop a prancing white warhorse, dressed in full military regalia, brandishing a sword like a dare.

"That woman doesn't understand you," my dad says about my mom. "I understand, we're the same, you and I, I know you need Ham, know you'll die without Ham." We understand the Force, he reminds me, and my mother does not. The Force won't let anything happen to Ham, my father promises, but soon it's the week before we're supposed to leave for Oklahoma, our last World Championship ride. The horses are scheduled to trailer out ahead, and the truckers won't load Ham until my dad's account is settled.

"I'm so ashamed," he chokes when I call him from my mother's kitchen. "But I just can't." Can't afford the expenses, the flight, the time off from work. He's driving for FedEx, a job he's kept secret but confesses now, just to me. "This stays between us," he says and then, "I love you, Sport."

"Love you," I say. We hang up. My mother pays for Ham's journey and mine. We leave for World without my father and though it shames me, I am relieved to be alone.

Soon it is the morning of my last ride. My mother has hung for-sale flyers all over the show grounds, hoping a new girl's parents will purchase Ham, and trailer him away at the week's end. My father is home in the woods and hasn't called. So it's my last chance, my last morning like this, the barn still before the big lights, Ham napping on the cross ties as I braid. I finish, move around the stall, primping. I dress—breeches, button-up shirt and collar, gold horseshoe pin; boots, jacket, gloves—the same uniform as the girls I ride against. I pin a synthetic blond bun to the back of my head, wipe my face in a slivered mirror and see exhaustion in my skin. I have fought for the length of the girlhood I am leaving now. I bridle, mount, disappear; we ride from the barn, toward the Coliseum, the show, the announcer's final call.

At home in the woods my father checks the time. He walks to his office, emptied of everything but his father's chair. He sits, closes his eyes to visualize my ride. It's dark before he sees me, projected by his mind's eye, in among the girls and horses. I am riding, and it's just like we've imagined, every step ordained. My father breathes to bridge the space between us. I breathe. Ham's breath matches mine. We stop. We breathe together. Together we wait.

———————

The judges deliberate. The Coliseum stills. There will be another war, I know this now. Win or lose, my father will remain in the cavern on the hill, foreclosure notices on the door, last pieces of taxidermy gathering dust on the walls. Win or lose, he will love me from that distance.

The World Champion is—my number, my name, and Ham's. The crowd erupts, music rises. We move to the Winner's Circle like there's a force there, pulling us.

The photographer takes our picture. The ring stewards approach with a blanket of roses; they cover us in all that bloom. The ring goes dark and the spot comes up, makes a circle in which Ham and I are safe, blind beyond the yellow edge. And still I picture my father, waiting in his father's chair. His palms are open, his body is a beam. We are alive together in the light.

WE AREN'T CLOSE TO ANYWHERE

ROSEBUD BEN-ONI

> *Hrafnkel had one animal in his possession that*
> *he valued more than others . . . which he named*
> *Freyfaxi . . . He had such a love for this stallion that he*
> *made an oath to bring about the death of any man who*
> *rode it without his permission.*

—FROM *THE SAGA OF HRAFNKEL FREY'S GOÐI*

When I talk about Odin now, I'll never forget my then husband's reaction to my stories and poems that were sparked by that relationship, which was not an *affair* per se, but has come to feel like one. I've been reading poems about Odin in public for the last few years, until one night when my usually calm and collected Brian spit out after a reading: "I can't believe four years later, you are *still* this obsessed with someone you knew for less than a week!"

I remember seeing the candid frustration in his eyes, holding in my breath—and then bursting out in laughter on the 7 train, as we made our way from Manhattan back to Queens.

Lest you think I'm a terrible person, please realize that

he was more upset that I'd fallen in a big way for this Odin *because* I knew him for less than a week.

And it doesn't make it any less threatening to him that Odin is, in fact, a horse.

In 2016, I went to Iceland to ride horses. Or, rather, we went on a two-week trip that included a five-day horseback-riding excursion. There *were* other reasons we went to Iceland. According to scientists, the Northern Lights would be the brightest they'd been in a decade, as they occur on an eleven-year solar cycle. Brian and I also found a good deal online. Yet while he was focused on hiking, visiting the capital Reykjavik, eating fresh fish, and trying specialties like svið (sheep's head, which I tried) and puffin (which I declined), all I could think about were the horses.

Four years before this, three life-changing things happened in the summer of 2012: I'd had my first book accepted for publication; I'd met the person I was going to marry; and on one strange summer day that same year, I inexplicably lost my balance while visiting family in the Rio Grande Valley. I was petting my uncle's pit bull Tiger when I felt a numbness radiate through my left extremities. Tiger, whom I'd just met, sniffed my tingling arm and whined. Less than a month later, it happened again, on Queens Boulevard in New York City, my left side fully giving out this time as I was crossing the multilane road. This was only the beginning. In the years that have followed, tingling and numbness have beset the left side of my body, along with head pain, brain fog, and spine tenderness. I never knew when these episodes would happen, and they happened erratically, in various degrees of pain and frustration.

So, when I say I went to Iceland to ride horses, I did not know riding Icelandic horses would also return a part of me: a wilder, freer part of me. I now look back upon my years before I got sick, at my old self, my foolish, fearless self who once hiked the infamous eight-mile trek up to the summit of Old Rag in Virginia and scrambled boulders without any prior experience, who'd walked the Great Wall of China in flip-flops. A wash-and-wear self who was so witty in her unwise ways, who did not know love and peace until I married. I moved around the world, alone, went without knowing where I'd be tomorrow, not caring about tomorrow, made rash and reckless decisions based on passion and impulse that left me sleeping in strange places, sharing strange rooms with strange women I barely knew whom I'd share a couple of love-struck, lustful weeks with, only to watch them disappear just as abruptly.

How I lived then was how I wrote: no rules, all abandon. I mistook this approach for a sort of revolutionary candor. It was the central truth on which I'd staked my life, one filled with neither long-term happiness nor unhappiness. I hadn't wanted to acknowledge it was also a matter of survival, attempting to live by my own rules. I felt that I was very much alone in the world, and that self-assurance sustained me. A life of torn sundresses, matted hair, and long bus rides that broke night, leading me, always, to a frightening elsewhere that I had to face, and did face, with a sort of optimism that I owed nobody anything. Most days, I'd wake up thinking I had made it this far, and there was always further to go. I was always, in a sense, awake and ready, open to the world, with quick, deft moves of working circuitry.

And foolishly, I'd thought I could live that way forever.

That is when I fell ill and that hardwired self-reliance proved not so innate after all; quite literally, my wiring had become faulty.

Much of the poetry I began writing since my illness (and am still writing) reflects this struggle to remain that old fearless me (which I still struggle not to call the "real" me), while also acknowledging that illness had actually had a profound impact on my work. In poetry, I found I could move on the page in the ways I sometimes couldn't in real life—*new* ways I hadn't even considered before—and I could experiment with language and ways of expressing a thought that I couldn't in everyday speech. Whereas I was uncertain that I could get from Point A to Point B without losing my balance in real life, on the page, I let myself "stumble" and meander, willingly, less afraid of what might happen.

While I thankfully had not experienced any flare-ups for the five months leading to the trip, there were certain activities of which I've been told to proceed with caution—horseback riding being one of them.

At first, I did not think much of what it meant to climb up and onto the back of a horse and surrender control. I live in New York City, after all, and though I'd occasionally seen people riding horses in Prospect Park in Brooklyn, my love and knowledge of horses remained separate from my urban life. I can count the number of times I'd actually been on horseback prior to the Iceland trip—four, all during trips to an uncle's farm in the Rio Grande Valley as a young girl. Each time was on the same ancient, gentle Quarter Horse who moved with a timid, almost apologetic gait. Her name was Rosa; a former racing horse, she was brown with a dark mane and tail. Her previous owner had abused her, and my uncle had taken her in. He didn't allow anyone else to ride her but me, as I was small, and would not be, as my uncle reasoned, a burden of weight on her back. Most of the time I preferred to watch her graze, or sit by her side when she lay down on the grass in the sun. I would talk to her and stroke her head. Once she laid her head in my lap, and I refused to get up, even after my legs fell asleep, even after the sun set in the sky and I hushed my uncle and my parents, shooing them to go away, to let us be.

I remember the day that my straightforward uncle leaned over the fence and told me I had to say goodbye to Rosa because she was very sick and he knew she was tired and wanted to find peace. I remember I cried as he lifted me up so I could throw my arms around her neck. I remember she stood very still. I remember understanding, in that stillness, how broken she was, how the love and care from my uncle could not fix her, could not erase the damage done to her. And what I remember most is her milky, cloudy eye, which she was trying to communicate through—or rather, a haze that *I* wanted to pierce through, for her to see the unconditional love that I had for her alone.

A caballo regalado, no se le ven los dientes, my uncle was fond of saying.

Literally: A gifted horse, you don't check the teeth.

What he meant: *Be grateful for your time together.*

So, many years later, I went to Iceland to ride horses, thinking of what it meant to be grateful for time.

I went to Iceland broken, though I hid it well, though I was loved and cared for by a kind and thoughtful person who, upon learning I was ill on our second date, stayed over and took me to the hospital the next morning for my seven a.m. blood tests.

I went to Iceland to ride horses, not knowing if I was going to be able to ride horses, uncertain if I still had it in me: that stubborn wild streak that now gave flesh to ghosts, to the ghost I'd become. To this new, phantasmal self, one who's not always content to live on the page, who doesn't want to proceed with much caution in my life. One who wants to break the rules given by my very well-meaning physician, because if not now, when? What if my condition got worse? What if this was my last chance to *be grateful for time spent together* with the horses I'd always admired from afar? Those still somewhat-wild, not completely tamed horses of Iceland for whom I longed in a strange way I didn't completely understand myself?

Perhaps I would arrive at our five-day trip and fail halfway through, in a place where there was nothing but miles of wasteland, where there was no one but humans who eyed me from horseback. No phone reception to call a car that couldn't come anyway.

Despite all my worries, it was this very uncertainty—particularly the uncertainty of my future—that propelled me to move forward with the trip. I didn't want to waste any more time.

Still, I had no idea what I was getting into when we arrived at the small farm and two horses ran up to greet me at the fence. They eyed me curiously, without fear, and let me rub their long faces. For a moment I was overwhelmed: I'd never seen such friendliness, such trust, in horses before. And I certainly did not know just what I was getting into when, quite suddenly, something nudged me—rather *hard*—from behind.

Jolted, I turned around. There, standing in front of me, blowing air through his nostrils, was a chestnut brown stallion. He already had a bridle and saddle on, already outfitted for another rider. He turned his head slightly to the side, and his eye shone like wet earth, the same color as the patch of mud under my feet. I looked around our small group in confusion as more and more people were assigned horses. After a moment, one of the guides came over, gave a little chuckle as he ran his hands through the stallion's mane, which was quite tangled, and asked about my level of riding.

I wasn't sure how to answer. I believe most people feel strongly about horses: they either love them or they don't. They either trust them or they don't. And while I did know I loved horses, I did not know the Icelandic horse at all. My older self would say: *So what? Jump in.* But this new, more cautious self felt more hesitant about how to answer such a question.

I turned to the two other horses at the fence who'd greeted me—much more docilely than the stallion who'd shoved me—but they'd sauntered off. Meanwhile, the chestnut snorted softly and rubbed his head against my hand.

I haven't been with horses for years, I finally said.

Oh yeah? the guide said.

I've lived in New York City for a long time, I said.

Oh yeah, he said.

Maybe too long, I added.

Oh yeah, I see. So. I have the horse for you.

Oh yeah? I said, cringing a bit, accidentally imitating him.

Oh yeah.

Which one?

Excuse me, excuse me! A woman called out, approaching us. She glared at me. *Excuse me. You have my horse.* She had a thick German accent, and was about a head taller than me, with a muscular build. She reached for the chestnut's reins.

Oh yeah, said the guide to her. *This one, this is not your horse.*

You just gave me this horse, the woman said.

This horse, you don't want, he said. *He's not gentle, and he has a mind of his own.*

But you gave me this horse already, she began.

Some horses can change color according to season, he added. *He does not.*

Why does that matter? the woman said. *What are you saying?*

I'm saying, the guide went on, *I have another horse for you.*

We both looked at him. Then he said to me: *But you, miss, you, I feel you can trust him—*

I felt my heart pound in my chest: That's what it was, wasn't it? That I needed a new level of trust I hadn't before.

—because you'll have to.

At that, the woman shook her head, and muttered something in German under her breath.

There was some nervous laughter from our group. Brian

cleared his throat and gave me a look that said he's about to intervene.

His name is Odin, the guide added.

This Odin looked at me with his light eyes of wet terracotta and nuzzled my shoulder.

I did not know the great amount of faith I'd have to put into this seemingly affectionate horse—who also turned out to be one of the most headstrong and difficult horses of our group—when crossing steep, rocky inclines and through a winding river again and again, the latter of which left me soaked head to toe, grasping onto his neck, trusting him while steering him in the right direction. I did not know the amount of times Odin would attempt to break away from the herd, and nudge my leg as if to say all the conflicting thoughts in his head—*it will be okay, believe in me, let me do what I want*—the times in which I'd let him have his way, temporarily, and the times I'd have to guide him back onto the chosen path.

That I would hold him closely all the times the temperature suddenly dropped, when I felt a sliver of my balance coming back to me as it had once been.

That my ghost self, this new, patchwork self, perhaps, could be as protective as she was stubborn, my own unreliable and capricious guardian angel who was learning to live in limbo, unchecked, without my full gratitude that had gotten me this far, to an island country that is still very much in the midst of *becoming*.

At the time, all I could think was: *Okay. I think I can handle him.*

———————

When I think about the limitless possibilities of poetry, I think of Iceland. A young island, less than twenty million years old, one that is still rising, erupting, splintering. What began as a subaqueous mountain created by volcanic activity along the Mid-Atlantic Ridge in the Atlantic Ocean is now an island that remains volcanic. From its numerous volcanoes to the black sand beach in the village of Vík í Mýrdal, its fields of lava and ash, its craters, pumice mountains, and columnar trap rock of basalt produced from slow-cooling magma, Iceland can appear volatile at times, and rather visibly: one can actually see the deep ridge between the two tectonic plates of Eurasia and North America, as they continue to pull apart from each other. Iceland is a continuous poem-in-progress in this way, one that shows all its drafts and revisions, a cacophony of poetry as multiverse, always moving, revealing landscapes as living things independent of our own existence. Still, our choices affect its future greatly: due to climate change, the country itself is literally rising, as its numerous glaciers might very well entirely melt away, leaving many to wonder what Iceland will become without its ice.

According to geology and paleogeography researchers like Dr. Christopher Scotese at the Field Museum in Chicago and the marine geophysicist Roy Livermore, it is also likely that one day, in a distant future, all landmass will become a supercontinent again, resulting in hypothetical, massive formations called Pangaea Proxima or Novopangaea. I wonder how humankind will fare on these supercontinents.

Iceland was largely uninhabitable when the first Norse settlers arrived. Without the horses, they and their subsequent generations in Iceland would not have survived; in the late ninth century, the Vikings brought over horses, mainly those of Germanic descent, and through a process of both preferred breeding and natural selection, only the strongest, most resilient horses survived the harsh Icelandic landscape. Only one percent of the country is "under arable cultivation," and the survival of the people in Iceland depended largely on the survival of their horses. Up until 1904, there were no cars in all of the country, so the horse was both the primary means of transportation and a vital source of labor.

Icelandic horses are *tough*, for sure, with their double coats and rock-star shaggy mane and tails; they have a big-boned yet sturdy build. But they are equally known to be affectionate, sure-footed, and less skittish than other breeds, since there are no predators that are a threat to them. Icelandic horses also have five different gaits: the usual walk, trot, and gallop, in addition to the tölt (running walk) and the skeið (flying pace), the latter of which requires both legs on the same side touching the ground at the same time. Not all Icelandic horses can perform the skeið, but the ones that can are very highly prized; their feet glide above the rockiest of terrain, at up to speeds of 30 mph. Perhaps it's through the conduit of the skeið that humans can achieve the gift of flight, suspended in that very limitless state of possibility that one might call poetry.

The Icelandic horse is one of the purest breeds in existence, yet, like the pristine nature of their home island, this is both because of, and despite, human choices and preferences. In order to protect them from diseases, even riding equip-

ment one brings *into* Iceland must be new or it will be con-
fiscated. Icelandic law also forbids the import of horses into
the country—of *any* breed—and those Icelandic horses who
leave are not allowed to return.

Ever.

While I've yet to complete it, I've been reading *The Sagas of
Icelanders* on and off for years. Many of the early Norse settlers went to Iceland because they would not bend to the will
of the kings, particularly Harald the Fair-Hair, who according
to *Egil's Saga*, revered his poets above "all his followers . . .
held [them] in the highest regard, and let them sit on the
bench opposite his high seat." Of course this regard came at
a price, which was personal and political freedom, and was
for some of these Viking raiders the one value they would not
give up.

Iceland then began as a place of both exile and freedom,
as the early settlers prospered from its waters rich in fish and
seals. They built farms. They also had to contend with the
large percentage of uninhabitable land, including the aus-
tere and barren Highlands where outlaws sought asylum. The
country's early governing years had no executive power and
no written laws, but an outdoor assembly, attended by the
most successful of farmers, gathered to settle disputes. It is
said that Icelanders today value their freedom above all else.
The country has never had its own national armed forces and
has resisted membership in the European Union.

So the fact that if a human chooses to take an Icelandic
horse out of the country, the horse is in a sense *banished* from

this very particular homeland, which is like nowhere else in the world, and seems both fitting and impossibly cruel.

Are there Icelandic horses right now, whether on a farm in Vermont or Norway, that are birthing their subsequent generations on foreign lands with the knowledge their foals will never see their homeland, because humans—another species who once depended on *them* for nearly a millennium—made a law to protect the purity of their species? Does an Icelandic horse ever think, *yes, this is for the good of the species*? Do they ever stop to consider us, those fragile two-legged creatures who do not, as they do, have the double coats and stamina to survive the winters, and yet are deciding their genetic fate?

What will the future look like, when the supercontinent happens, and the Icelandic horses meet other breeds, meet relatives they lost generations back?

What will they remember?

And will humankind still be there, determined to uphold its laws of return and national borders? Will it be our foolish belief in human exceptionalism—in this new, united landmass that brings together all—that will keep us so far away from everything?

The first day was the toughest. Our group headed down a steep incline into one of many rivers and streams. Riding an Icelandic horse, I never felt the rough, rocky ground or the inclines; it truly is different than riding any other breed. On foot, I most certainly would have stumbled, even at a slow pace, or had to crouch down to pull myself up, resorting to some sort of awkward four-limbed climbing. But the ride went smoothly—that is, until Odin chose to drink from the river as we were crossing. His head went down and the rein tightened, bringing me down with him. The first time this happened, I felt my feet fall out of the stirrups. I grasped onto him for dear life.

Aw, how sweet, someone said as they calmly passed us, *look how she's embracing him.*

Odin then put his whole head in the water, shaking it furiously. I felt myself falling. Suddenly he stopped and righted himself up. He nudged my left leg, rubbing his muzzle against me slowly, as if in apology. Or rather, as if to appease me, aware of his charms. I patted his head gently, and pulled him to the right and whispered: *Let's go*. He then broke into a quick trot. My feet were still not in the stirrups, and I held him closely as we raced up the incline and came to level ground. I pulled him to stop. He stopped. I struggled to find my balance and secure my position in the saddle. A guide came over and scolded: *Odin! Not today!*

In response to this rebuke, Odin and I glided past everyone, all the other riders, and then the head guide. We'd already been warned to not let our horses do this, that such an action is a challenge to authority. I pulled him to stop. I felt the collective judging eyes of the other riders. Including Brian's. I felt embarrassed and helpless. I could not control my horse. I leaned down to warn him, grumbling in his ear: *Cut it out. Or I'm going to switch horses with someone.*

I suspected that Odin understood my frustration perfectly.

I also suspected he knew switching horses was impossible at this point.

So we repeated this several times: he'd keep trying to pass everyone, and each time he did it, I pulled him to stop. Eventually, I stopped talking to him. I sat there, stoically on top of him. Each time, when he'd then nudge my left leg, I didn't react.

It takes trees a long time to grow here, a guide said, sidling up next to me. *Because of the climate. Same with training these horses. They are tough, but have minds of their own.*

After an hour of this stopping and starting, Odin seemingly tried to make amends. There were about forty riders total, with four guides, but no matter where we were on the trail, Odin would bring me back continuously to Brian's horse, so we could ride together. At one point, the same guide pulled up alongside us, and said: *He's smart. He knows this man is your mate.* Then he added: *He's being good. Don't expect it to last too long.*

But still, this is the horse for me, right? I sighed.

Do you want to change horses? Brian called out to us.

My guide shook his head at both of us.

You have to trust him, miss, he said.

Do you trust him? I asked him.

I don't know, the guide said. *I've actually never ridden him. He's very difficult, and usually a curmudgeon.*

Odin then took the opportunity to surge off again, but this time I pulled him back quickly, continuously pulling and falling against him, until he stopped.

See? the guide said, pulling up beside me.

See what? Brian called over, catching up to us again, sounding more frustrated than I was.

When we dismounted for lunch, Odin went to eat grass with the other horses. Since we weren't allowed to take photos on horseback—for good reason, our safety—as frustrated as I was, I wanted to get a quick photo with my horse, to remember our first day together. Brian and I had talked about living in the moment—taking a break from our phones, computers, and social media—but I couldn't help it. I asked him to hold on to Odin's reins while I took a photo of him from afar. Odin responded by trotting over toward me, dragging Brian along with him, and nuzzling my phone.

I then made the mistake of taking a running head start to attempt to get this photo.

Because Odin runs faster than I do. Much, much faster. Brian was nearly sailing through the air.

I realized Odin was not going to cooperate. He snorted loudly, and nuzzled the phone in my hand, as Brian handed me the reins. Odin was fascinated with the phone as I flipped the camera our way to take a selfie. I couldn't get his whole head in the photo. Odin leaned in very close, so close, if I turned my head, I would have gotten a mouth full of his hair.

This horse *posed* for a picture.

But also, once again he got his way, and we took the kind of photo he wanted.

Afterward, Odin settled down on the ground. He was, for a moment, very serene.

The guide said: *You can sit next to him. Odin likes you. He likes your company.*

Brian was more cautious and said: *We aren't close to anywhere.*

We aren't close to anywhere? I said. *What do you mean?
What if he kicks you by accident, or gets on top of you—
What?*

I MEAN if he ROLLS on top of you—

I looked at Brian. He was looking at Odin, frowning. I
thought I must be imagining things, but it did seem that he
was a bit, well, *jealous*. But he also wore a familiar worried expression, one that I'd seen before, one that asked what would
I do if I lost my balance or all the feeling on my left side. What
would I do—or rather, what would *we* do—if that pain found
me in a remote place on a remote island, far from my neurol-
ogist and our families—our sense of home and security.

And yet, this was exactly what I needed: to stop being so
cautious.

To give in, not foolishly, but for something, someone, who
was important to me.

Even if, at that time, that *someone* was a horse I'd known
for less than a day. Or maybe *because* he'd known me for less
than a day, but could sense things about my nature in a very
real, nonhuman way. Maybe because Odin didn't seem too
worried that I'd hurt myself. Maybe because he was, in the
words of the guide, *not gentle* and had *a mind of his own*. After
all, *he* first nudged *me*, and I had to be up for the challenge of
dealing with a very willful horse that also expressed a tenderness toward me. I knew that he wouldn't, in the end, hurt me.

All I could say to Brian was that I'd be careful. I tried to
be, anyway. I lowered myself gently to the ground, and patted
Odin's head and then his neck. He closed his eyes. I lay against
him, gently, and then more assuredly. I heard him breathing.

It sounded like a great hall filling with oxygen, with a world, with whatever is the opposite of emptiness.

For a moment I was in the suspension of a horse's breath.

Somewhere, the restless ghost quieted her anxious desire to make herself known, that parts of my old self were still *here*.

Somewhere, she wrestled herself into an idea of just *becoming* by *being*.

A caballo regalado, no se le ven los dientes.

Be grateful for your time together.

Odin and I were silent as can be.

We were, just for a moment, the two of us, and indeed, not close to anything, or anywhere, but each other.

At that moment, I placed my trust in that.

I still do.

That first day, in late afternoon, we crossed a river on the way to our destination for that evening, a homestead where we'd spend the night. Odin and I had fallen behind again, and were the last ones to cross. It began to sleet and rain at the same time. I remember when we first entered the river I studied the horses who'd just crossed, their hindquarters soaked and shaking off torrents of cold water. Odin took me nearly straight down a slope, the feeling of falling returning as I straightened my spine and held on to him with my legs. We entered the river, and when we reached halfway, he stopped.

I spoke to him gently, pulled at the reins, tried to push him onward with the forward motion of my body. But he didn't move.

Across the river, the group had already gone on toward

camp, and only Brian and the guide remained, waiting for me on the other side of the bank. I saw Brian turn around and look at us. I realized I was sitting on a very still horse in the middle of a fast-moving river, with water up to my waist.

I felt panic rising up in my chest.

Though Icelandic horses can swim quite well, and can survive treading the cold, glacial waters, what if the horse suddenly bucked, my foot got caught in the stirrup and my head was submerged until I drowned? What if I got hypothermia? Or what if this were some strange test of nature, some cruel twist to the idea of "trust," one animal (him) testing another (me) to see just how far the other would go before giving into fear and surrendering, my dignity waving some inner white flag?

The guide called out to Odin, and then called to me, and then to both us in one quick breath, almost as if our names had become entangled, a new kind of expletive—*Rosebodin!*

Odin remained unfazed.

I felt him breathing calmly, his body contracting and expanding between my legs.

Like he was waiting.

Waiting for me to see just *where* we were: beyond the misty sleet falling around us, beyond the middle of a freezing river. Beyond the body, that which I felt I could not control. *My* body in which I'd lost faith and trust, as I sat so still upon another so certain, so solid and powerful, that which had been simultaneously challenging and protecting me. A rowdy, graceless, and rather rude horse that, in the end, chose me, perhaps because he sensed how broken and fearful I'd become not only of my body, but of life itself.

It was in that moment, as the guide began to turn his horse around to reenter the river, that I became fully present.

Something kicked inside me, that old instinct to think fast on my feet, that which got me through scrambling boulders on my own and walking home late at night in East Jerusalem by myself on badly lit roads, where no railing protected me from sharp and sudden cliffs.

Only now I wasn't completely alone.

But Odin wasn't here to protect me; he was here to encourage me. Although he knew the way, although he could brave these waters for much longer than I ever could, he was waiting for me to stop being afraid, to lead us both out of the river.

Looking back, as Brian once told me, it all happened very quickly, so quickly that by the time Odin and I reached the side of the other bank, the guide had not yet even entered the water. But at that time, it seemed like a lifetime of cold water seeping through my clothes, until the chill reached the ghost, *my* ghost. Something kicked inside me; it was the old me. The first me. I turned myself over to Odin as he turned himself over to me. I pulled the reins to the right, and kicked him gently to move. My entire lower body was in the river. I could do this. I leaned into him, and then lifted myself up. The water was moving fast but we were faster. We were reactive. We were volatile.

What I have been is not yet lost.

I often wonder what will happen not only to myself in the future, but also to humankind. And I have come to believe if we all disappear, I will not mourn this world. Because I am quite certain these horses—horses that thrive in the uninhabitable—will inherit these continents that someday will be pushed together. Because I am quite certain that Odin knows his thousand-year history, in a way that does not require words.

I would not weep if the future means no speech like our speech, no language, no poetry as we have defined it. For there is a poetry beyond us, beyond humans. If in future eras no one knows what it means to "read" and our books go unread, turn to dust, to earth, while the über-horse lives on, breaking our laws, disregarding our borders, and running alongside any horse of their choosing, I would not think this world lost.

Perhaps we have to overcome what makes us human to understand the rest of the world around us. Perhaps part of that is letting go of what we think is best for all living things. Perhaps part of that is forgetting language, forgetting the self as what language alone makes. Yet in order to convey these experiences and ideas what else can I use?

I wish I could send every poet and writer a horse for their dreams, for when I returned home from the trip and was reading *The Sagas of Icelanders*, I came across the name of the horse I'd been given—Odin. I learned he was named after the patron god of poets.

When Odin and I came out of the river, we soon joined our group dismounting at the homestead. My horse shook the water off his body and grazed while Brian and I showered in hot water heated by geothermal power. Then, having changed clothes, Brian and I sat bundled up in a field looking at the stars, waiting for our guide who'd take us to "chase" the Northern Lights.

Before we turned in for the night, I said good night to Odin, and that I'd see him in the morning. I laid my head against his still-damp coat. I could hear him breathing. I still do. I will always remember that deep chamber of being, the rising and falling of him. I left part of myself there, within him, in that soft, baked earth of an eye, his wet terra-cotta eye, that weary eye, if weary were the sum total of his world. *Trust me*, says the eye. *Trust*.

If we could entrust this planet to only one kind, I'd stake my life on the horses.

May they inherit the earth.

CONTRIBUTORS

C. MORGAN BABST is a native of New Orleans. She studied writing at Yale and NYU, and her essays and short fiction have appeared in the *Washington Post*, *Saveur*, the *Oxford American*, *Guernica*, *Garden and Gun*, and the *Harvard Review*, among others. Her debut novel, *The Floating World*, was named one of the best books of 2017 by *Kirkus*, Amazon, *Southern Living*, and the *Dallas Morning News* and was a *New York Times* Editors' Choice.

Born to a Mexican mother and Jewish father, **ROSEBUD BEN-ONI** is the winner of the 2019 Alice James Award for *If This Is the Age We End Discovery* (Alice James Books, 2021) and the author of *turn around, BRXGHT XYXS* (Get Fresh Books, 2019). Her chapbook *20 Atomic Sonnets*, which appears in *Black Warrior Review* (2020), is part of a larger future project called *The Atomic Sonnets*, which she began in 2019 in honor of the periodic table's 150th birthday. She is a recipient of the 2014 NYFA Fellowship in Poetry and a 2013 CantoMundo Fellow. Her work appears in *POETRY*, the *American Poetry Review*, the *Academy of American Poets' Poem-a-Day*, *Poetry Society of America (PSA)*, the *Poetry Review* (UK), *Tin House*, *Guernica*, *Black Warrior Review*, *TriQuarterly*, *Prairie Schooner*, *Electric Literature*, and *Hayden's Ferry Review*, among others. In 2017, her poem "Poet Wrestling with Angels in the Dark" was commissioned by the National September 11 Memorial & Museum in New York City, and published by the *Kenyon Review Online*. Her poem "Dancing with Kiko on the Moon" was recently featured in Tracy K. Smith's *The Slowdown*. She's part of the 2018 QUEENSBOUND project, founded by KC Trommer, and took part in the Onassis Foundation's 2020 ENTER exhibition. She writes weekly for the *Kenyon Review* blog.

BRAUDIE BLAIS-BILLIE is a Brooklyn-based writer hailing from the Seminole Tribe of Florida's Hollywood reservation. Focusing on the intersection of Indigenous issues, music, and culture, her work has appeared in publications like *Pitchfork*, *i-D*, *Glamour*, *Billboard*, and more. Braudie is the founder and editor of *indige•zine*, a digital and print platform for Indigenous art, identity, and resistance.

ADRIENNE CELT is the author of three novels, including *End of the World House*, which is forthcoming from Simon & Schuster; *Invitation to a Bonfire*, which was named a 2018 Indie Next Pick and a best book of the year by the *Financial Times*; and *The Daughters*, which won the 2015 PEN Southwest Book Award for Fiction and was shortlisted for the 2016 Crawford Award. A collection of her comics, *Apocalypse How? An Existential Bestiary*, was published by New Michigan Press in 2016. She is the recipient of an O. Henry Prize and a Glenna Luschei Prize, and residencies at Jentel, Ragdale, Lighthouse Works, and the Willapa Bay AiR, among other honors. Her work has appeared in *McSweeney's Quarterly*, *Strange Hori-*

zons, *Zyzzyva*, *Epoch*, the *Kenyon Review*, the *Paris Review Daily*, and many other places. She lives in Tucson, Arizona.

SARAH ENELOW-SNYDER is a freelance writer who grew up in Spicewood, Texas, and has lived in the New York area for the last fifteen years. She has bylines in the *New York Times*, the *Washington Post*, *Condé Nast Traveler*, *High Country News*, theGrio, and many others.

NUR NASREEN IBRAHIM is a writer, journalist, and producer. Her fiction and nonfiction have been included in anthologies and collections from Catapult, Hachette India, Platypus Press, *The Aleph Review*, *Salmagundi* magazine, and more. She is a two-time finalist of the Salam Award for Imaginative Fiction.

CARMEN MARIA MACHADO is the author of the bestselling memoir *In the Dream House* and the award-winning short story collection *Her Body and Other Parties*. She has been a finalist for the National Book Award and the winner of the Bard Fiction Prize, the Lambda Literary Award for Lesbian Fiction, the Lambda Literary Award for LGBTQ Nonfiction, the Brooklyn Public Library Literature Prize, the Shirley Jackson Award, and the National Book Critics Circle's John Leonard Prize. In 2018, the *New York Times* listed *Her Body and Other Parties* as a member of "The New Vanguard," one of "15 remarkable books by women that are shaping the way we read and write fiction in the 21st century." Her essays, fiction, and criticism have appeared in the *New Yorker*, the

New York Times, Granta, Vogue, This American Life, Harper's Bazaar, Tin House, McSweeney's Quarterly Concern, The Believer, Guernica, Best American Science Fiction & Fantasy, Best American Nonrequired Reading, and elsewhere. She holds an MFA from the Iowa Writers' Workshop and has been awarded fellowships and residencies from the Guggenheim Foundation, Yaddo, Hedgebrook, and the Millay Colony for the Arts. She lives in Philadelphia and is the Abrams Artist-in-Residence at the University of Pennsylvania.

T KIRA MAHEALANI MADDEN is a writer, photographer, and amateur magician. She is the author of *Long Live the Tribe of Fatherless Girls* (Bloomsbury).

ALEX MARZANO-LESNEVICH is the author of *The Fact of a Body: A Murder and a Memoir*, which received a Lambda Literary Award, the Chautauqua Prize, the Grand Prix des Lectrices de ELLE, the Prix des libraires du Québec, and the Prix France Inter-JDD, an award for one book of any genre in the world. It has been translated into ten languages. Their next book, *Both and Neither*, is a genre-and-gender-bending work of memoir, history, cultural analysis, trans reimaginings, and international road trip about life beyond the binary. It is forthcoming from Doubleday. An essay adapted from the book appears in *Best American Essays* 2020.

COURTNEY MAUM is the author of the novels *Costalegre, I Am Having So Much Fun Here Without You*, and *Touch*; the popular guidebook *Before and After the Book Deal: A Writer's*

Guide to Finishing, Publishing, Promoting, and Surviving Your First Book; and the forthcoming memoir *The Year of the Horses*. Courtney's nonfiction has been widely published in such outlets as the *New York Times* and *Modern Loss*, and her short story "This Is Not Your Fault" was turned into an Audible Original at Amazon. Courtney is the founder of the collaborative retreat program The Cabins, and she has a creativity advice newsletter you can sign up for at CourtneyMaum.com.

ALLIE ROWBOTTOM is the author of the memoir *Jell-O Girls*, a *New York Times* Editors' Choice selection. Her essays and short fiction can be found in *Vanity Fair, Salon, Best American Essays*, and elsewhere. She is the recipient of fellowships and awards from Disquiet, Summer Literary Seminars, Inprint, and Tin House. Allie holds a PhD from the University of Houston and an MFA from CalArts and lives in LA.

MAGGIE SHIPSTEAD is the *New York Times* bestselling author of the novels *Great Circle, Astonish Me*, and *Seating Arrangements*, winner of the Dylan Thomas Prize and the *L.A. Times* Book Prize for First Fiction. She is a graduate of the Iowa Writers' Workshop, a former Wallace Stegner Fellow at Stanford, and the recipient of a fellowship from the National Endowment for the Arts.

JANE SMILEY's newest novel is *Perestroika in Paris*, about a horse who escapes from Auteuil Racecourse and lives in Paris through the winter. She has written many (too many) books about horses, and this is her favorite.

LAURA MAYLENE WALTER's debut novel, *Body of Stars*, was published by Dutton in March 2021. Her writing has appeared in *Poets & Writers*, the *Kenyon Review*, *The Sun*, the *Masters Review*, *Ninth Letter*, and many other publications. She has been a Tin House Scholar, a recipient of the Ohioana Library Association's Walter Rumsey Marvin Grant, and a writer-in-residence at Yaddo, the Chautauqua Institution, and Art Omi: Writers. She lives in Cleveland, where her Breyer horses are still packed away in the attic.

ACKNOWLEDGMENTS

Foremost, thank you to the courageous and talented writers who contributed essays; without you this book wouldn't exist. Thank you to my agent, Sarah Bowlin, my editor, Sarah Stein, and the team at Harper Perennial for making *Horse Girls* possible. Thank you to my brilliant colleagues at Electric Literature, past and present, particularly Preety Sidhu for her thorough fact-check, and Jess Zimmerman for saying, "Someone should publish a book called *Horse Girls*, and it should probably be you." Thank you to my first readers and advisers: Julie Buntin, Libby Flores, Leigh Newman, Steph Opitz, and Matt Sumell; your support and guidance encourage me to write and to take risks.

To Susie Beale, Katie Rinda, and Doug King, thank you for teaching me to ride. To all the horses and ponies I have ridden, or at least as many as I can remember—Silver, Gunny, Tennessee, Lucy, Johnny, Paylon, Charlie, JP, Yankee, Dave, Moose, Woody, and Reagan—thank you for sharing your freedom.

I am endlessly grateful to my parents, Carl and Karin, for their everlasting support; to my sister, Noori, for her lifelong friendship; and to my husband, Paul, for one million years.

ABOUT THE EDITOR

HALIMAH MARCUS's short stories and essays have appeared in *One Story*, *BOMB*, the *Literary Review*, Amazon Original Stories, the *Out There* podcast, *Indiana Review*, *Gulf Coast*, the *Southampton Review*, and elsewhere. She is the executive director of Electric Literature, an innovative digital publishing nonprofit, and the editor in chief of its weekly fiction magazine, *Recommended Reading*, which she cofounded. She has an MFA from Brooklyn College and lives in the Catskill region of New York.